SpringerBriefs in Applied Sciences and Technology

PoliMI SpringerBriefs

For further volumes:
http://www.springer.com/series/11159
http://www.polimi.it

Pierluigi Freni · Eleonora Marina Botta
Luca Randazzo · Paolo Ariano

Innovative Hand Exoskeleton Design for Extravehicular Activities in Space

Pierluigi Freni
Applied Science and Technology
Polytechnic University of Turin
Torino
Italy

Eleonora Marina Botta
Mechanical Engineering
McGill University
Montreal
Canada

Luca Randazzo
Control and Computer Engineering
Polytechnic University of Turin
Torino
Italy

Paolo Ariano
Center for Space Human Robotics
Italian Institute of Technology
Torino
Italy

ISSN 2282-2577   ISSN 2282-2585   (electronic)
ISBN 978-3-319-03957-2   ISBN 978-3-319-03958-9   (eBook)
DOI 10.1007/978-3-319-03958-9
Springer Cham Heidelberg New York Dordrecht London

Library of Congress Control Number: 2014940157

Printed on acid-free paper

Springer is part of Springer Science+Business Media (www.springer.com)

Preface

The work presented in this monograph is the final result of the multidisciplinary project entitled COmponents for SPAce Robotics (COSPAR), developed in the period 2010–2012 within the context of the Alta Scuola Politecnica and in cooperation with the Istituto Italiano di Tecnologia (IIT). The project team was constituted by M.Sc. students selected among the most talented ones of both Politecnico di Torino and Politecnico di Milano, in particular:

- *Eleonora Marina Botta*, Space Engineering, Politecnico di Milano.
- *Pierluigi Freni*, Materials Engineering, Politecnico di Torino.
- *Ippazio Martella*, Computer Engineering, Politecnico di Torino.
- *Federico Radici*, Biomedical Engineering, Politecnico di Milano.
- *Luca Randazzo*, Mechatronics Engineering, Politecnico di Torino.
- *Roberto Rossi*, Mechanical Engineering, Politecnico di Milano.

All the components of the team gave their personal and fundamental contribution to the present work and those who are not cited among the authors are indicated at the beginning of the chapter they contributed the most.

The Alta Scuola Politecnica

The Alta Scuola Politecnica (ASP) is a school for young talents founded by Politecnico di Torino and Politecnico di Milano in 2004. The ASP program is profoundly multidisciplinary and is focused on innovation. 150 students per year are chosen for this advanced program, which runs parallel to their M.Sc. courses. During courses, they enrich their knowledge of innovation processes and the socioeconomic context. In their multidisciplinary projects, they learn to develop a whole project working in a diverse team cooperating with industries, research, and governmental institutions.

COmponents for SPAce Robotics

The Components for Space Robotics (COSPAR) project aims at developing new technologies that could support space exploration missions. In the next decades, planetary exploration will play an important role in directing the global technological development and will provide at the same time an extensive application and testing field for many innovative technologies. In support of exploration missions, new space systems shall be developed requiring research on many new technologies involving robotics, automation, bioengineering, artificial intelligence, and nanotechnology skills.

One of the main problems during Extra-Vehicular Activities (EVA) is the astronaut's hand fatigue: the pressurized suit limits the astronaut's mobility and, in particular, the hands' dexterity, force and endurance. Astronauts would benefit from a device able to mitigate hand fatigue during EVAs without hindering any movement, allowing them to accomplish their tasks more comfortably and for a longer time. In this project, the team's approach was to research a possible preliminary technological solution for a lightweight hand exoskeleton to be embedded in the astronaut's glove. The difficulty of the project was increased by the complexity of the human hand, in terms of degrees of freedom and working space, and by the extreme environment in which the exoskeleton will be employed. The team focused on highly innovative and high-risk solutions designed from scratch for a glove based on soft robotics concepts, employing Electroactive Polymers (EAPs). The design solution is an innovative response that makes it possible to foresee the subsequent stages of prototyping and development.

Acknowledgments

We are glad to express our gratitude to all the people and institutions involved in the COSPAR project, that made this experience possible for us, gave us their support and enabled the results attainment. We would like to thank the ASP board for offering us the opportunity to work on this project, within a highly specialized context, and for the continuous support both on organization and content aspects. A very special thank you goes to Paolo Ariano of IIT, for his continuous help and collaboration along the whole project development and for the several precious inputs and ideas he shared with us.

We want also express sincere gratitude to Paolo Nespoli for offering us an irreplaceable occasion to understand the astronauts' needs and the problems they have to face during EVAs, which brought us to a much higher level of awareness and led us to the final concept design able to meet the final users' priorities.

Special thanks go to Loredana Bessone for giving us the possibility to participate in a visit to the European Astronauts Center, to Riccardo Bosca for the guided tour of EAC's facilities and to Herve Stevenin for sharing precious information about astronauts' training.

Sincere thanks are due to Emanuele Tracino, Project Manager of Thales Alenia Space, with whom we had extremely constructive confrontations and who particularly helped us in the understanding of radiation hazards in EVAs.

Conflict of interest. The authors report no conflict of interest.

Contents

Abbreviations

CAD	Computer-Aided Design
CIR	Instant Centre of Rotation
CMC	Carpo-Metacarpal
CNT	Carbon Nanotubes
DC	Direct Current
DE	Dielectric Elastomer
DIP	Distal Interphalangeal Joint
DoF	Degree of Freedom
EAC	European Astronaut Centre
EAP	Electroactive Polymer
ECLSS	Environmental Control and Life Support System
EEG	Electroencephalography
EKG	Electrocardiogram
EMG	Electromyography
EMU	Extravehicular Mobility Unit
EVA	Extra-Vehicular Activities
GTO	Geostationary Transfer Orbit
IPMC	Ionic Polymer-Metal Composites
ISS	International Space Station
IIT	Italian Institute of Technology
IVA	Intra-Vehicular Activities
LCE	Liquid Crystal Elastomers
LEO	Low Earth Orbit
LET	Linear Energy Transfers
MEBA	Multilayer Elongation and Bending Actuator
MCP	Metacarpophalangeal Joint
MMG	Mechanomyography
MU	Motor Units
MUAP	Motor Units Active Potential
MVC	Maximum Voluntary Contraction
NASA	National Aeronautics and Space Administration
NEA	Near Earth Asteroid
PANI	Polyaniline
PIP	Proximal Interphalangeal Joint

PVDF   Polyvinylidene Difluoride
RMS   Root Mean Square
SAW   Surface Acoustic Waves
SEE   Single Event Effects
SEM   Soft Extra Muscle
sEMG   Surface Electromyography
SEL   Single Event Latchups
SEU   Single Event Upsets
TrFE   Trifluoroethylene
VDF   Vinylidene Fluoride

Chapter 1
Introduction

Human activities in space are fundamental for many exploration missions and include Extra-Vehicular Activities (EVAs), which represent some of the most challenging, hard and dangerous duties. The extravehicular pressurized suit imposes severe limitations to the astronaut's mobility, impacting particularly on dexterity, force and endurance of the hands. A device able to mitigate hand fatigue during EVAs would be a significant improvement for astronauts, allowing them to accomplish their tasks more efficiently, more comfortably and for a longer time. The present work presents the preliminary design of a hand exoskeleton that could enhance crewmembers' performance during these operations. *Soft robotics* was the leitmotiv throughout all the project development, together with the attention to the stakeholders' and users' requirements. In addition to literature review, a user driven design approach was followed to identify the requirements and some mockups were built in order to determine the most important hand degrees of freedom. Experiments assessed the effectiveness and applicability of surface electromyography (sEMG) and mechanomyography (MMG) for signal acquisition and the material was selected with a multi-criteria analysis. Finally, after the definition of some design alternatives, the final concept solution was chosen.

Mankind has always been wondering about space, but space exploration has been just a dream until the 20th century. Thanks to the technological progress, traveling outside the Earth's atmosphere became possible about 50 years ago, giving start to the so-called Space Race. The presence of humans in space missions is fundamental for many technical aspects and has a meaningful outcome in terms of social and cultural impact on public opinion. Crewmembers play a crucial role in carrying on the space mission tasks and the Extra-Vehicular Activities (EVAs) represent some of the most challenging, hard and dangerous duties. The extravehicular pressurized suit imposes severe limitations to the astronaut's mobility, impacting particularly on dexterity, force and endurance of the hands. One of the main problems limiting the overall duration of a spacewalk is the astronaut's hand fatigue. A device able to mitigate hand fatigue during EVAs would be a significant improvement for astronauts, allowing them to accomplish their tasks more efficiently, more comfortably and for a longer time. The ability of moving effectively

P. Freni et al., *Innovative Hand Exoskeleton Design for Extravehicular Activities in Space*, PoliMI SpringerBriefs,
DOI: 10.1007/978-3-319-03958-9_1, © The Author(s) 2014

wearing the space suit will become increasingly important in the future, with space missions aiming at exploring other planets.

The aim of the project was to design a hand exoskeleton that could enhance crewmembers' performance during these operations. This is a preliminary study of a possible technological solution able to reduce the fatigue of the astronaut's hands, while avoiding hindering its natural movements. The challenge was the realization of one or more prototypes of a lightweight hand exoskeleton to be embedded in the astronaut's glove, in order to overcome the stiffness of the pressurized suit. Both the high complexity of the human hand, in terms of degrees of freedom and working space, and the extreme environment in which the exoskeleton will have to work create a series of different constraints increasing the complexity of the project. The team focused on innovative and high-risk solutions designed from scratch for a soft glove. *Soft robotics* was the leitmotiv throughout all the project development and was chosen as opposed to the more traditional *hard robotics* approach. Therefore, the research about the state of the art, as well as the selection and concept design activities, were carried out within a *soft* framework. In the light of the highly innovative field the team worked within, the first phase of the project was devoted to assess the state of the art of the available exoskeleton solutions. At the same time information and data were collected in order to have an overview about the technologies representing the edge of development in sensors, control and actuation systems. To perform this research the group was divided in two sub-teams: one focused on acquisition and control, the other on materials and actuation. Bibliographic research and data collection ran in parallel to the other activities throughout the whole project development: heavy contributions came from the participation to conferences, congresses, international fairs and meetings with stakeholders.

Among the preliminary activities was the identification of stakeholders' and users' requirements; in this respect a user driven design approach was followed. Furthermore, the study of hand and fingers movements was performed, building some mock-ups, in order to determine the most important degrees of freedom to be implemented in the solution. Some direct experiments were carried out to assess the effectiveness and applicability of different signal acquisition technologies, such as surface electromyography (sEMG) and mechanomyography (MMG). After gathering data and information, the second phase of the project development was devoted to the material selection, with a multi-criteria analysis, the definition of some design alternatives and the selection of a final concept solution.

Chapter 2
Users' Requirements

Abstract This chapter deals with the identification of the requirements that drive the design of the hand exoskeleton to help crewmembers during Extra-Vehicular Activities (EVAs). The requirements were identified by means of literature review as well as interviews to users and stakeholders. After an introduction to EVAs, the method followed for the design is presented. The space environment and the main characteristics of spacesuits and gloves are reviewed, focusing in particular on the condition of the hands. The fatigue problem of arms and hands during EVAs is explained and some peculiarities of the training that astronauts undergo to prepare to this type of activities in space are stressed. Since one of the most important problems for materials and electronics in space is radiation, the total dose that has to be withstood in the typical International Space Station (ISS) environment is then estimated. Different requirements also come from the EVA spacesuit equipment, as well as from safety and cost considerations. A discussion on kinematics and dynamics follows, in which the main hand movements needed for EVAs and the different joints of the fingers are discussed. Finally, a table summarizes the identified requirements, which drive the design of the hand exoskeleton.

2.1 EVAs

Spacewalks are a fundamental component in space exploration: they are used to install new equipment, perform repairs, set up and control experiments. To understand the importance of such activities, it is enough to consider that almost 1,000 h of Extra-Vehicular Activities have been needed to build and maintain the International Space Station (ISS) since 1998.

When they work in space, astronauts wear protective suits, made of many different layers, which help keep them at the right temperature, enable them to breathe and protect them from harmful radiation (Fig. 2.1). Because of conditions in space and inside the spacesuit, movements are particularly difficult and require

P. Freni et al., *Innovative Hand Exoskeleton Design for Extravehicular Activities in Space*, PoliMI SpringerBriefs,
DOI: 10.1007/978-3-319-03958-9_2, © The Author(s) 2014

Fig. 2.1 Russian cosmonaut Anton Shkaplerov, Expedition 30 flight engineer, in Russian Orlan spacesuit, while performing Extra-Vehicular Activities for the outfitting of the International Space Station (NASA Courtesy)

high resistance to fatigue. The aim of the present work is to develop a hand exoskeleton in order to help astronauts during EVAs.

2.2 The Method

The group worked in the direction of a user driven design, trying to understand deeply the needs of the direct users, the astronauts, and to translate these into requirements. Among the activities carried out in order to acquire a better perspective on the subject of EVAs are two important events: an interview to Paolo Nespoli and a visit to the European Astronaut Centre (EAC) in Cologne.

Paolo Nespoli was the first Italian astronaut staying on board the ISS and has a vast experience in coordinating EVAs. On April 20th, 2012 he was in Milan to deliver a presentation to Politecnico di Milano students, and the ASP board organized a restricted meeting for ASP students working on space-related projects. During the meeting, the team had the opportunity to ask Nespoli some questions about the conditions that astronauts face when engaged in EVAs and the main obstacles that they need to overcome.

Another important event was the visit to the European Astronaut Centre (EAC) in Cologne. During the visit, the team could gain knowledge of the methods and activities used to train European astronauts before they leave the Earth.

Moreover, during a first visit to IIT laboratories in Turin, some team components had the opportunity of trying a Russian astronaut glove on. Although the piece of equipment was not recent, the experience helped the team understand that the EVA glove fabric causes noticeable resistance and fatigue in performing repetitive movements, even neglecting the stringent problem of pressure.

Different space-oriented companies were also involved to some extent: the team visited Selex-Galileo laboratories near Milan and had some constructive exchanges with Emanuele Tracino, a Project Manager of Thales Alenia Space.

The knowledge the group acquired in the above-mentioned occasions is reported in the following part of this chapter, integrated by the results of literature research and personal considerations.

2.3 Space Environment

Space environment is particularly hazardous; a number of factors cause the degradation of materials and components and entail health issues. Atomic oxygen present in the ionosphere reacts with polymers and some metals, eroding the external layers; vacuum causes the evaporation of external volatile substances, which can contaminate other exposed surfaces; space debris hit exposed walls and provoke critical damages; temperatures vary of hundreds of degrees from daylight to shadow. Radiation is the most relevant problem outside the absorbing protection of the atmosphere and magnetosphere and leads to the degradation of materials, depending on the quantity of absorbed energy, called dose. Ionizing radiation (due to electrons, protons, ions, x-rays and γ-rays) hits external surfaces and penetrates for a certain depth in the material; this degrades vital elements as electric components and solar cells.

The orbits of interest for EVAs are currently the ISS' and, potentially in the near future, Near Earth Asteroid (NEA)'s. The ISS orbits at a mean altitude of about 330 km, with an inclination of 51.6° above the Equator; this is a Low Earth Orbit (LEO). Objects in LEO are partially protected from radiation thanks to the Earth's magnetosphere, but the South Atlantic Anomaly (a zone where radiation inner belt is nearer to the Earth's surface) exposes astronauts and devices to harsher conditions. NEAs do not travel at a constant distance from the Earth; during missions on asteroids the astronauts would practically experience deep space conditions, characterized by solar winds and Galactic Cosmic Rays. In this case working conditions during EVAs would be more severe, particularly as far as radiation is concerned.

2.4 Spacesuits and Gloves

As Paolo Nespoli said during the meeting with ASP students, *a spacesuit for a spacewalk is essentially a portable spacecraft that astronauts put on before exiting the ISS*. An Environmental Control and Life Support System (ECLSS) provides the astronaut with breathable air (100 % oxygen at low pressure) and other consumables and takes care of cooling. Two different EVA suits are onboard the ISS: the American and the Russian, both with dedicated gloves and helmets. Nespoli stressed that there is a big difference between Russian and American gloves: Russian gloves are tough and do not allow much dexterity, while American ones are better for mobility since they are often customized, even at phalanx level.

Any spacesuit is made of multi-layered fabric in order to counter environmental hazards. Anyway, the main problem with astronauts' fatigue is not the resistance of the spacesuit itself, but the pressure inside it. On Earth we breathe air at a pressure of 1 atm (14.7 psi). This value is too high for a spacesuit, since pressure outside is almost null and mechanical efforts would be too high to be tolerated by the fabric and to allow any movements. Decreased suit pressures enhance crewmember flexibility and dexterity, reducing space suit operating forces and pressure loads; on the other hand, they make oxygenation more difficult and increase the risk of decompression sickness. Several studies have shown that if the gas inside the spacesuit is pure oxygen, then the pressure can be lowered to about one third of an atmosphere (4.3 psi), which is a reasonable compromise between the competing constraints (Patrick et al. 2010). Nonetheless, the effort to perform some movements is noticeable, especially in bending fingers to grasp objects.

The temperature inside the glove, and the spacesuit in general, is a direct consequence of the temperature outside, which in space changes very rapidly and with considerable ΔTs (-123 to $+232$ °C) (Patrick et al. 2010). To counter temperature gradients, the suits are equipped with a liquid refrigerating system that cools the astronaut when the external temperature is too high; however, there is no particular way to put in heat in case the astronaut is cold. Typical temperatures in the spacesuit are 10–27 °C, but some extreme temperatures are often experienced. Sometimes modules get so hot that they cannot be touched, and astronauts have to stop working and wait in a shadow area in order to cool down their gloves. Cases of frostbites have been experienced too; therefore, American gloves' tips are now equipped with little resistors that can be switched on when necessary.

Sweat is a big issue during spacewalks. The duty of the 91.5 m narrow tubes of the Liquid Cooling and Ventilation Garment is to make water circulate around the crewmember's body and remove extra heat. The vents in the garment draw sweat away, and this is recycled in the cooling system. Oxygen helps the refrigeration, too: it is pulled in at the wrists and ankles to help with circulation within the spacesuit. On the hands, however, no cooling system is provided. Under the glove astronauts wear a special layer which is supposed to weak away sweat from the fingers, but, according to Nespoli's experience, this is not sufficient to avoid people returning from EVAs with fingers *as if they had been in water for hours*. Sweat is a

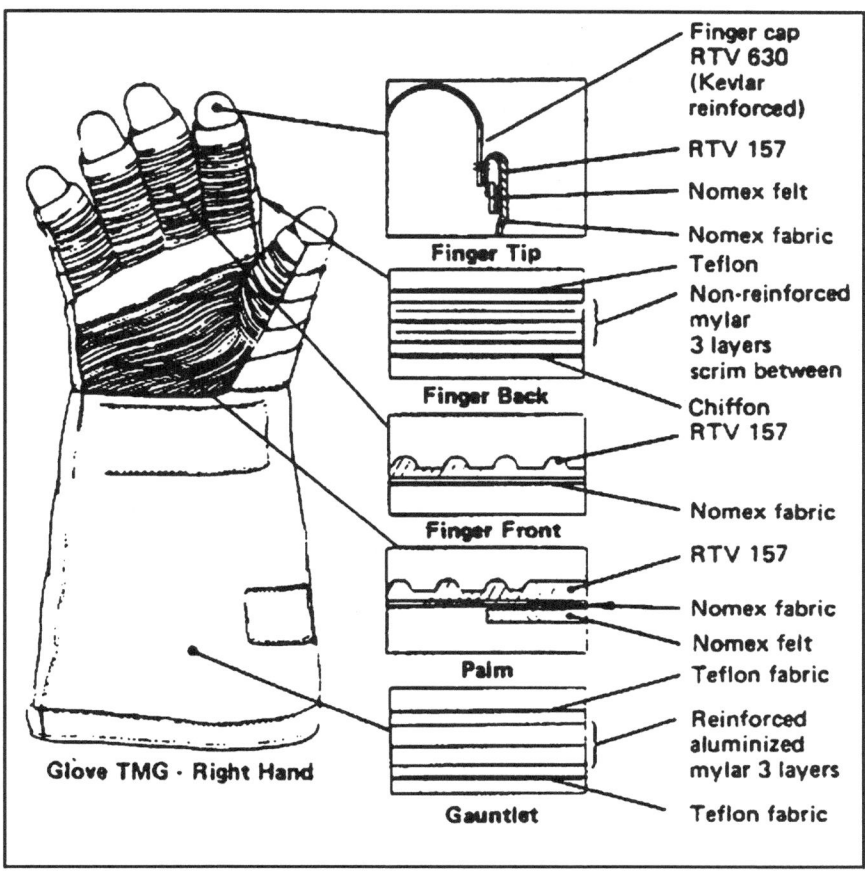

Fig. 2.2 Shuttle EVA glove fabric structure (NASA 2011)

critical point for the team's research: it has a detrimental impact on the accuracy
and performance of certain technologies (for example electromyography). In case
a device is located on the hands' skin, it would get wet and lose in performance; it
would be better to put it on the forearm.

An example attesting the efforts in the design of EVA spacesuit gloves is the
STS EMU glove Thermal Micrometeoroid Garment depicted in Fig. 2.2. It is
straightforward to see that this variable cross section protecting covering is the
result of a complex design aimed at optimizing hand mobility, abrasion resistance
and thermal protection. Different materials are used in the same part of the glove,
to grant different types of protection, and different cross section layouts charac-
terize parts subject to different types of stressing conditions. A few peculiarities
are worth being underlined: RTV resists to abrasion and enhances gripping and
tactility, while providing protection; Mylar reflects solar radiation; Teflon grants
abrasion and thermal protection.

NASA Space Suit Glove Design Considerations recognize the need for pressurized gloves that can be worn for extended periods of time without undue discomfort, and that allow firm grasp retention of handholds, switches, tools, etc., for short periods of time without hand fatigue (NASA 2011). Disregard the above-mentioned design efforts, STS EMU gloves considerably degrade tactile proficiency compared to bare hand operations. Nonetheless, sufficient dexterity to operate many standard handles, knobs, toggle switches and buttons is assured.

2.5 Fatigue Problems During EVAs

In the ISS program, many EVAs have been performed: spacewalkers were charged with building and are continuously repairing the International Space Station. *Spacewalkers manipulated elements up to 9,000 kg, relocated and installed large replacement parts, captured and repaired failed satellites, and performed surgical-like repairs of delicate solar arrays, rotating joints, and sensitive Orbiter Thermal Protection System components* (Patrick et al. 2010). These tasks do not typically require a lot of force, but high resistance to fatigue. Even simple operations in space become complex because of the peculiar working conditions. First of all, with the EVA glove, the fingers are fixed in a neutral position and any motion to change the fingers or hand position is a struggle against the spacesuit, which acts as a balloon. Lack of gravity precludes leverage exploitation: as an example, on the Earth, torque applied with a screwdriver is opposed by a counter-torque passively generated by the weight of the user; in space, it must be countered actively exerting a muscular force in the opposing direction. Moreover, while working, the astronauts need to stay always attached to some handle bar with one hand, for safety reasons; this leads to further effort for one limb.

The preparation for an EVA is really complicated: it begins with the decompression inside the ISS even the day before the dive. Because of the huge effort in preparation that these procedures require, during an EVA astronauts have to accomplish as many tasks as possible. Duration of a spacewalk is limited by consumables (water, electrical power, oxygen regeneration) and is about 7 or 8 h, even though this depends on individual metabolic rates. Despite pains taken to ensure performance and comfort with customized gloves, working long hours in what fundamentally is a pressurized balloon can be grueling on the hands and body, astronauts say. EVA tasks require many finger/hand/arm motions for several hours, leading to significant fatigue. Sometimes the effort is such that the heartbeat frequency is too high and the astronaut has to stop his/her activities for some time.

2.6 EVA Training

Strength capabilities are situation-specific: within the same person, they are influenced by body position and restraints at the worksite location. It is therefore necessary to verify by means of simulation that EVA crewmember potential responds to any required EVA task. The EAC in Cologne houses some replicas of the different parts of the ISS, where astronauts are trained on how to perform the different experiments, how to safely use the instruments and, most importantly, how to deal with dangerous situations both during IVAs (Intra-Vehicular Activities) and EVAs. Test engineers of the EAC are capable of injecting different types of faults in the equipment and teach the astronauts the correct behavior to contain and solve various problems.

During the visit to such facility, some components of the team and of IIT attended a meeting with Herve Stevenin, Head of NBF Operations and EVA Training Unit and Eurocom Team Leader, who focused on the description of the training for EVAs. He reported that he EAC is equipped with a large swimming pool, where conditions of low or absent gravity can be reproduced. In the pool, mockups of different portions of the ISS are immersed so that trainees can simulate the actual movements they need to perform during EVAs.

The main problems that astronauts have to face during EVAs are the limitations to the freedom of movement. Current spacesuits designs do not allow to move the arms upwards (like when climbing a ladder), so that the only type of movement which can be performed for displacement is lateral strafing. Moreover, astronauts have to work always with one hand gripping the handholds and the other performing the various tasks; leaving the handholds just for few seconds could be fatal. For this reason, tools are designed to be used with one hand and spacesuits are equipped with snap-hooks that can be used to secure the astronauts to the station. A further problem encountered by astronauts during EVAs is the limited field of view; the helmet does not allow any movement of the neck, leading to considerably limited visibility. In this context, astronauts are forced to memorize the position of every tool in their belt (they cannot directly see them) and to plan each and every step before moving.

Even though the pool training is not useful for reproducing many conditions experienced in space, it leads to a dramatic decrease in the time required to complete the final stages of astronauts training at NASA training center. This kind of program prepares astronauts both physically and mentally, creating through exercise the ability to withstand typical workloads and providing them with a deep knowledge of the ISS environment. Anyway, the above-mentioned problems of fatigue persist; for this reason, a device that could help endure this problem would be convenient and useful.

2.7 Radiation

One of the most stringent requirements for materials and electronics in particular is radiation. Electronics, in fact, is subject to damages caused by Total Dose (total energy per unit mass deposited in a material by ionizing radiation) and Single Event Effects (SEE- produced when single, ionized particles interact with electronic devices, changing their electrical state).

2.7.1 EVA Suit Glove Shielding

A NASA report (Cucinotta et al. 2003) provides some evaluation of water-equivalent shield thicknesses for the different parts of spacesuits. In the following, both suits available on the ISS for EVAs are detailed. The thinnest part of the glove is the finger covering, since it has to allow sufficient dexterity so that astronauts can grab handholds and manipulate tools and station components; this represents the worst case condition for any device that may be inserted in the glove. The equivalent shielding values are summarized in Table 2.1 (a: Average of right hand, dorsal entrance, index and ring fingers; b: Average of right hand dorsal entrance, middle and distal phalanx).

A NASA specification (NASA et al. 1994) provides information on the worst case radiation environment in the ISS orbit, for different sources, according to equivalent Aluminum shielding thickness. The above-reported glove thickness can be transformed in equivalent Aluminum by means of the following formula:

$$X_{Al} = \frac{A_{Al}}{Z_{Al}} \cdot \frac{Z_{H_2O}}{A_{H_2O}} \cdot X_{H_2O}$$

where A is the atomic weight and Z the atomic number (or effective atomic number). The resulting values are reported in Table 2.2.

According to the aforementioned specification, this protection leads to a total dose of $6.397 \cdot 10^3$ rads Si (where 1 rad Si $= 100$ ergs per gram of material) in one year, if the device is shielded by the spacesuit's glove only. EVAs are not performed during solar flares, whose radiation can be fatal to humans; therefore SEEs do not need to be taken into account for the glove thickness alone. However, these calculations are not able to give an exact requirement for the material and electronics. In fact, this would be valid if astronauts were continuously performing EVAs; actually, EVAs are still sporadic and any device is stored in the ISS for the rest of the time.

Table 2.1 Water equivalent shielding values for EMU and Orlan-M gloves

	Thickness ($g \cdot cm^{-2} H_2O$)
EMU Glove (a)	0.198
Orlan-M Glove (b)	0.198

Table 2.2 Aluminum equivalent shielding values for EMU and Orlan-M gloves

	Thickness ($g \cdot cm^{-2} Al$)
EMU glove (a)	0.171
Orlan-M glove (b)	0.171

2.7.2 ISS Shielding

According to Boeder et al. (2005), the shielding thickness of the ISS ranges from 10 to 100 g/cm^2 Al. According to the same NASA specification used before, the resulting total dose if the device was always stored in the ISS under a 10 g/cm^2 Al thickness protection would be of 25 rads Si.

An estimate of Total Dose can be obtained by weighting the results of this and the previous paragraph on the time spent inside and outside the ISS. The typical number of EVAs is 12/year; remembering that an EVA lasts about 8 h, the time spent in EVA each year is straightforward. During this amount of time the device is shielded by the glove alone, while during the rest of the time it is stored on the ISS. As a result, the weighted average total dose is of about 95.1 rads Si.

Influence of solar flares radiation cannot be totally discarded while the device is stored onboard the ISS: important Linear Energy Transfers (LET) can cause Single Event Upsets (SEU—state change of a memory element) and Single Event Latch-ups (SEL—potentially destructive parasitic currents). SEU are not permanent errors and can be countered by providing a correction algorithm. SEL, on the other hand, must absolutely be avoided: space-qualified LET immune electronic components need to be used in the final design. Real-time monitoring of solar activity is also necessary to avoid performing EVAs during solar flares, which are particularly dangerous for humans.

As a final consideration, during EVAs, astronauts will normally be carrying a tool or gripping a handhold. This offers protection to the ventral side; if we want to protect the electronics from radiation, an interesting possibility would be locating it on the palm.

2.8 EMU Suit Battery

For Shuttle Extravehicular Mobility Unit (EMU) spacesuit, electric power during EVAs is provided by a rechargeable silver/zinc battery supplying 16.8 V and containing 26.6 Ah of energy when fully charged. Recharge time is of 19 h, onboard the ISS.

2.9 Safety

NASA General EVA Safety Design Requirements state, among the others, that electrical current limiting devices shall be provided to eliminate all potential ignition sources within any oxygen-enriched atmosphere of the life support system and pressure suit and the EVA crewmember shall be protected against electric voltage shocks from inadvertent grounding of electric circuits and from electrical discharge resulting from static charge buildup (Abramov et al. 1994). Energy storage or conversion devices that may cause a short-circuit and a discharge rate capable of igniting materials are not allowed unless some current limiting is provided. Therefore, to comply with NASA requirements, the device must be provided with current limiting equipment and high voltage shocks must be prevented.

A crewmember performing an EVA is continuously subject to a real-time physiological monitoring of hearth rate, EKG signal, suit pressure, O_2 consumption, partial pressure, CO_2 pressure and radiation exposure. These monitored parameters allow the knowledge of the health status of the crewmember, the possibility to face any emergency and the management of tasks planning according to fatigue status. This monitoring capability may be complemented by specific stress or fatigue indexes as EMG or MMG, to identify the occurrence of localized fatigue, for example.

2.10 Space Market and Stakeholders

Cost is a fundamental limitation to space missions. The cost of space transportation is evaluated in *cost per pound to orbit*: putting a kg of payload in LEO costs thousands of dollars; the cost more than doubles for higher orbits (like GTO—Geostationary Transfer Orbit). As an example, according to Futron Corporation (2002), one kilogram of payload sent in LEO by means of Soyuz Launcher cost US$5,357. Being space launches so expensive, light materials are preferred both for spacecrafts and accessories; this project is an attempt to propose an innovative concept of exoskeletons, solving the weight problem through the use of alternative light materials. During the meeting with Mr. Stevenin, IIT was allowed to present the concept of hand exoskeleton based on the research performed together with the team. After the presentation, Mr. Stevenin stated that the project looked promising.

2.11 Kinematic and Dynamics Analysis and Considerations

During the visit to the EAC in Cologne, the group acquired a better understanding of what kind of activities astronauts are asked to perform during EVAs and which are the most important hand movements that they need to make. Because

Fig. 2.3 The eight features
of movements (Abolfathi
2008)

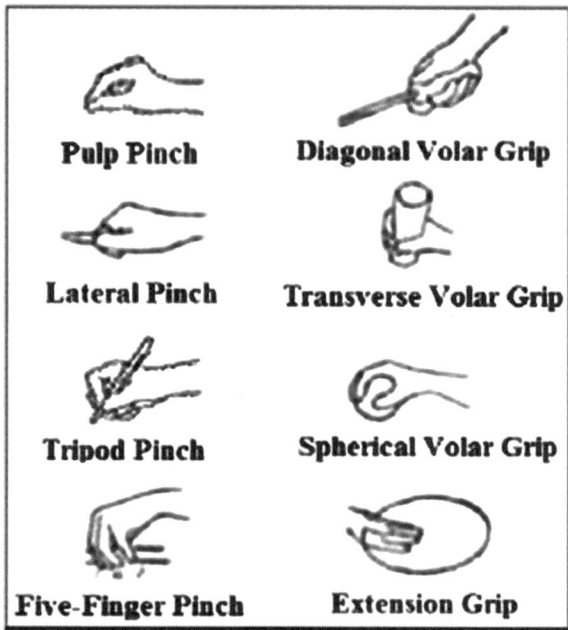

astronauts need to be safely hooked to the spacecraft at all times, handles are
available all over the hull for the operators to grasp. This grasping movement must
be performed for a very long time and is also necessary for the utilization of most
of the tools employed during EVAs. The second most important hand movement
commonly performed during EVAs is pinching, occurring whenever the astronauts
need to interact with smaller objects or knobs.

From a more general perspective, we can underline how the interviews and the
studies narrowed the field of action. In fact, the hand has 23 DoF, which would be
absolutely unfeasible to be developed. Nevertheless, different studies conducted on
human movements allow reducing the complexity of the problem: it is demon-
strated that humans only perform eight features of movements, identified in
Fig. 2.3 (Abolfathi 2008). Sensing and control systems can be highly simplified by
employing this assumption. Moreover, the information obtained through interviews stresses which features should be identified and actuated more robustly and
with better performances.

Having identified the needed features, the following step is to define more
specifically the kinematics and dynamics of the model in order to produce detailed
specifications for the project. Researchers agree on identifying three different kinds
of joints in the hand (Kapandji 1998). 1-DoF joints can be modeled as hinges,
whereas 2-DoF joints take the shape of a saddle, as the metacarpal joint, or of a
condyloid joint, as the CMC (carpo-metacarpal joint) of the thumb. These joints
are recognizable from Fig. 2.4.

Fig. 2.4 Scheme of the hand's joints (reproduced with permission of Van der Smagt et al. 2009)

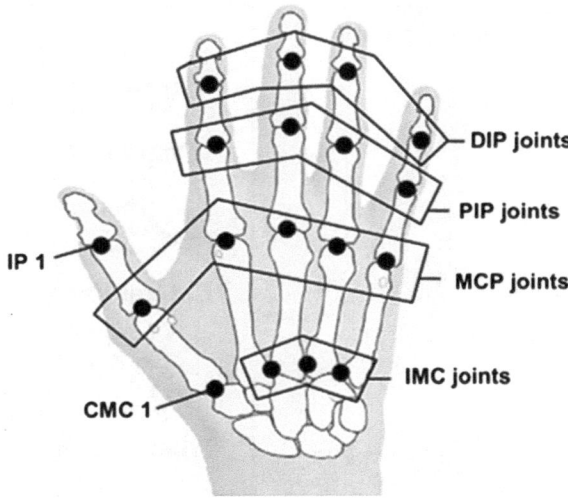

Fig. 2.5 Illustration of the three joints of a finger (Abolfathi 2008)

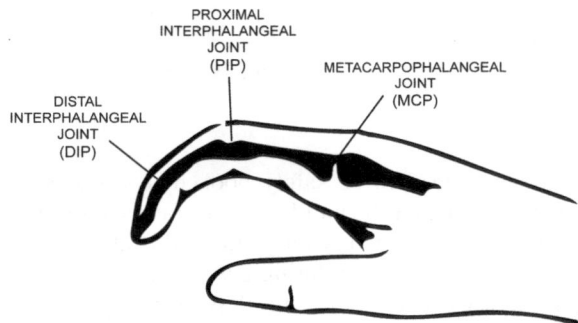

Studies conducted by Patrick Van der Smagt et al. (2009) show the numerical values of the center and the axes of rotation of each joint of the hand. As a consequence, once the correct way of actuation of the three kinds of joints understood and the particular ones we want to actuate chosen, it would be straightforward to apply to these the *soft actuation* as we have conceived it.

The differences between the three kinds of joints add further complexity to the problem and discriminates the mechanically built structure to the smart material structure from the kinematics adherence perspective. A mechanically built structure can hardly follow a moving Instant Centre of Rotation trajectory without employing a bulky structure. Smart soft material design can easily deal with kinematics complexity just by changing the punctual response of the material within the space of work.

The method used consists in addressing and completely characterizing a specific part of the dynamical problem to be able to extend it to the other joints and fingers. As a consequence, the team focused on the dynamics of the finger.

The three joints reported in Fig. 2.5 are 1-DoF (DIP and PIP) and 2-DoF (MCP). A dynamical model of the finger was built to better define the required forces and to draw numerical specifications. The model will be presented in the Chap. 5, where the dynamic aspects are more organically addressed.

2.12 Requirements Table

The main requirements discussed in the previous paragraphs are summarized in the following table. Division is made between functional, safety, interface and environmental requirements.

Requirement identifier	Statement
	Functional requirements
RD_F1	Pinching and grasping with thumb, index finger and middle finger must be augmented
RD_F2	Total weight shall be minimized
RD_F3	The device shall fit inside the astronaut glove
RD_F4	The structure shall not hamper the physiological dexterity
RD_F5	The power consumption shall be minimized
	Safety requirements
RD_S1	The device shall not damage the skin of the user
RD_S2	In case of breakdown, the device shall not be an ignition source
RD_S3	In case of breakdown, the device shall not constitute electric shock hazard
RD_S4	In case of breakdown, the structure shall not hamper the physiological dexterity
RD_S5	The device shall not cause hazards to the mechanical structure of the hand
	Interface requirements
RD_I1	The user interface shall be seamless
RD_I2	The device must work always according to the user will
	Environmental requirements
RD_E1	The device shall operate in a range of temperature between 0 and 40 °C
RD_E2	The device shall operate in moist environment
RD_E3	The device shall survive the launch environment
RD_E4	The device shall withstand the radiation environment of the ISS
RD_E5	The device shall operate in presence of LEO radiation environment
RD_E6	The device shall operate at the pressure of 4.3 psi
RD_E7	The device shall operate in a pure oxygen atmosphere

References

Abolfathi PP (2008) Development of an instrumented and powered exoskeleton for the rehabilitation of the hand. University of Sydney, Sydney

Abramov IP, McBarron JW, Severin GI, Whitsett CE (1994) Chapter 14: individual system for Crewmwmber life support and extracehicular activity. In: Space Biology and Medicine II—Life Support and Habitability, US/Russian Publication AIAA-Nauka Press

ASI; Canadian Space Agency; ESA; NASA; NASDA (1994) Space Station Ionizing Radiation Design Environment—International Space Station Alpha. NASA Space Station Program Office, Johnson Space Flight Center, Houston, Texas

Boeder PA, Koonts SL, Pankop C, Reddell B (2005) The Ionizing Radiation Environment on the International Space Station: performance vs. expectations for avionics and materials. NASA Johnson Space Center, Houston

Corporation Futron (2002) Space transportation costs: trends in price per pound to orbit 1990–2000. Bethesda, Maryland

Cucinotta FA, Saganti PB, Semones E, Zapp N (2003) Chapter 5: a comparison of model calculation and measurement of absorbed dose for proton irradiation. In: Radiation Protection Studies Of International Space Station extravehicular activity space suits, pp. 97–103

Kapandji A (1998) The physiology of the joints. Churchill Livingstone, New York

NASA (2011) Chapter 14: extravehicular activity. In: NASA Space flight human-system stadard volume 1. Washington, NASA

Patrick N, Kosmo J, Locke J, Trevino L, Trevino R (2010) Extravehicular activity operations and advancements. In: Wings in orbit: Scientific and Engineering Legacies of the Space Shuttle 1971-2010, NASA/SP-2010-3409, pp. 110–129

Van der Smagt P, Grebenstein M, Urbanek H, Fligge N, Strohmayr M, Stillfried G, Parrish J, Gustus A (2009) Robotics of human movements. J Physiol 103(3–5):119–132

Chapter 3
State of the Art

Abstract In this chapter we wrap up our literature investigation, pointing out the key focuses and requirements to be considered. We will go through our evaluations upon actuators, sensors, control systems and we will present some designs examples. Although literature is rich in studies on limbs exoskeletons aimed to several target applications, (such as rehabilitation, function restoring and virtual reality), the stringent requisites of EVA make most of the actuation and sensing solutions unsuitable for our device. Safety reasons and pressurization of the space suit force to avoid modern actuation technologies (i.e. air muscles), and focus on traditional motors such as piezoelectric ones. For what concerns sensors and control systems, most of the currently developed exoskeletons use physiological signals and are based on Electroencefalography (EEG) and on Electromyograpy (EMG). In the latter case, a big advantage is due to the fact that input signals are picked up directly from the motor units involved in the hand control. In the next pages we also present several examples of exoskeletons, including a pinching device for rehabilitation based on a cable mechanism, one finger device based on pulleys-lever structure, the four bar mechanism, a tendon system glove for function restoring and the EVA K-Glove from NASA. Finally, a couple of examples of feedback devices for medical applications are presented, integrating a variety of sensors, such as accelerometers, ultrasound, flow, pressure and vibration sensors, heat infrared camera, gyroscopes.

Several studies concerning the development of various limbs exoskeletons have been carried out by research institutes, taking into account different applications and needs. Considering the hand movement, in particular, a couple of macro-categories can be considered; besides Extra Vehicular Activities, rehabilitation and function restoring in patients with impaired or lost hand functionalities represent some of the main needs that lead the research in this field.

Unfortunately, although literature is rich in exoskeleton prototypes and examples, most of them are totally unsuitable to space applications: they have been

---

Chapter written with the contribution of Federico Radici (Politecnico of Milano).

P. Freni et al., *Innovative Hand Exoskeleton Design for Extravehicular Activities in Space*, PoliMI SpringerBriefs, DOI: 10.1007/978-3-319-03958-9_3, © The Author(s) 2014

developed within the area of orthotic devices and virtual reality, with no strict concern about size and bulk of sensors, actuators and control units, which do not represent a problem in these endeavors. EVA applications present a series of peculiar and stringent requisites that heavily narrow the possibilities spectrum, due both to the astronauts' needs, hurdled dexterity and hard operating conditions, and to the space environment.

This second category includes serious limitations to the technology to be used, because of thermal and electrical disturbances such as electromagnetic radiation, plasma and geomagnetic storms (Favetto et al. 2010). At the same time, the necessity of embedding all the devices within the space suits carries some particularly stringent size limitations, which lead both to the minimization of the dimensions and to the placement of the actuators away from the hand itself (i.e. forearm). Finally, considering the high number of the hand's degrees of freedom (23 for the hand and 4 each finger), a compromise should be considered between the variety of movements to be allowed and the technological possibilities of control. In virtual reality applications, for instance, joint movements are often limited and only mono-directional movements are usually considered, which do not present problems in the haptic feedback and motion simulation, but would affect heavily both rehabilitation and space applications (Wege and Günter 2005).

Nonetheless, among the structures studied in the literature, many devices were concretely taken into consideration for the project, as far as the actuation, the sensing system and the concept itself are concerned.

3.1 Actuators and Sensors

Actuators are probably the most problematic component of the whole design. Even though different modern technologies are available, such as high pressure liquid-based actuators and high pressure air-based actuators (i.e. air muscles), the strict limitations of EVA application force to exclude these technologies and to focus on traditional motors, including electric motors, such as brushless or stepper, and piezoelectric motors (Favetto et al. 2010). In general, high pressure gases or liquids should be avoided for safety reasons and due to the pressurization of the space suit. At the same time, energy saving needs and magnetic field external contaminations must be considered.

One interesting solution is offered by piezoelectric motors, which are highly controllable and tunable and whose extremely reduced size (i.e. millimeters–centimeters) could be promising for the glove embedded placement. Despite the reduced dimensions, they allow high forces thanks to the dynamics deriving from the assembly, geometry and functioning of the piezoelectric components placed into the motor (Henderson 2007). The system is electrically controlled and based on the acoustic or ultrasonic vibration of piezoelectric materials under an electrical stimulation.

Several studied configurations of the components allow to convert this vibration into movements in the desired directions. Among the available configurations to translate the vibrations into the aimed movement (either linear or rotational) is the coupling of three crystals groups operating to push the stator. One Motive crystal, responsible for the motion, is connected to the motor case or the stator, while two Locking components are placed around it. Other configurations are based on surface acoustic waves (SAW), and can promote both translation and rotation. A drive technique normally associated to the trademark Squiggle®, instead, is based on the bounding of piezoelectric actuators to a threaded nut coupled with an internal threaded screw. The deformation of the nut into a precise orbit leads to the linear motion of the inner screw through a direct drive mechanism (New Scale Technologies n.d.). Finally, current piezoelectric technology allows an extremely uniform and consistent response to the electric field stimulation, and for high frequency of the steps, which can reach 5 MHz, with a linear speed upper limit of about 800 mm/s (Favetto et al. 2010) (Fig. 3.1).

Other types of actuation usually imply much bigger sizes and require alternative solutions. One approach that is considered is the use of Bowden cables in order to allow the displacement of the control unit and actuators from the hand itself (Wege and Günter 2005). Within this approach, the motors can be placed, for example, on the forearm, avoiding the mechanical hurdling of the hand and finger movements (Dean and Flores 2012).

The same kind of problems arises in the choice of sensors, which typically have to be positioned in the proximity of the source of signal to be acquired. Furthermore, the limitations due to the space environment have to be taken into account, once again, although a proper coating of the devices can help in protection and isolation (Favetto et al. 2010).

Among the possible sensing systems, some promising alternatives include strain gauge-based electrogoniometers, piezoresistive sensors, stretch/band sensors and EMG sensors. Piezoresistive sensors are based on the resistance variation of a silicon-based piezoelectric diaphragm due to a mechanical deformation and allow the acquisition of forces and pressures. Electrogoniometers and stetch/band sensors, instead, are based on the variation of resistance due to either an angle variation or a stretching and bending movement. They can both employ a Wheatstone bridge, which grants protection from the effects of thermal variations (Favetto et al. 2010).

3.2 Control Systems

Most of the currently developed exoskeletons are controlled using physiological signals. In this regard, the trend techniques are Electromyograpy (EMG), and Electroencefalography (EEG); they involve recording electrical signals from, respectively, the muscles and the brain. Both techniques can be invasive or not.

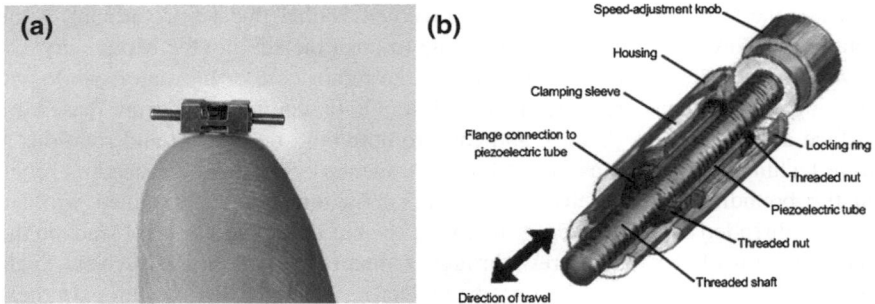

Fig. 3.1 a The picture shows qualitatively the size of a piezoelectric motor. This type of devices present extremely contained dimensions, in the order of millimeters. **b** A schematic of a Squiggle piezoelectric motor. The drive mechanism is based on the rotation of the screw inside a threaded nut which is deformed in a planned orbit by the bounded piezoelectric actuators; the piezoelectric vibration is therefore converted into the linear movement of the screw (reproduced with permission of New Scale Technologies Inc.)

3.2.1 EEGs

As far as brain signals are concerned, many case studies of invasive recording techniques have been proven to work really well (Hochberg et al. 2012; Velliste et al. 2008). Their main advantage is that the electrodes can be placed on the cerebral cortex portion that is directly involved in the motor control (Porro et al. 1996); it has been proved that, in invasive EEG experiments, the acquired signal quality is higher than in non-invasive cases (Ball et al. 2009). The problem with this technique is that it requires "intensive" surgery and, for sure, it would not be suitable to common users.

There are also many non-invasive techniques that allow to use skin-electrodes placed on the user's head (Emotiv). One of the main drawbacks of these devices is cross-talking (Boulenger et al. 2008): as electrodes are not immediately in contact with the motor-cortex, the signals they are recording comes from places on the brain that are related to different tasks than motion control. Additionally, the skin-electrode contact plays a fundamental role in order to have a good signal and motion-artifacts could also heavily affect the acquired signal (Friesen et al. 1990).

Both techniques, invasive and non-invasive, have shown that the user is required a lot of "concentration" during the experiments and this quickly leads to a tiredness state (Fig. 3.2 and Table 3.1).

3.2.2 EMGs

Most of the arguments related to EEGs are also valid for EMGs. They can be acquired both with invasive and non-invasive electrodes; in the case of EMGs, electrodes are usually placed as close as possible to the muscles to be detected. As

Fig. 3.2 The eMotiv headset is an example of commercial EEG devices (reproduced with permission of Emotiv)

Table 3.1 Pros and cons concerning the use of non-invasive EEGs

Pros	Cons
Lot of research being done on this topic	Lot of concentration is required from the user to produce "useful" signals
	Skin-electrode contact plays an important role

before, invasive techniques result in a more precise understanding of the muscle involved in the motion; this means that it would be easier to actuate the corresponding degree of motion on the exoskeleton. But, also in this case, invasive electrodes require not desirable surgical operations on the user. As for EEG signals, if non-invasive electrodes are used (surface EMG), there will be a lot of cross talking between the signals coming from the different electrodes, meaning that a certain effort is required to understand to which extent a particular signal set is related to certain muscles activation (Vuskovic and Li 1996) (Fig. 3.3 and Table 3.2).

In both cases (EEG and EMG), a lot of training is required for the user and a control algorithm is needed to understand how to map the signals acquired from the user with a corresponding input signal for the exoskeleton.

3.2.3 Other Solutions

Recent research has also shown new types of devices that could be used as sensors/feedback for the control system. They are a form of electronics that fairy precisely matches the skin's mechanical properties (Northwestern n.d.). These new devices could pave the way for sensors that monitor heart or brain/muscles activity without bulky equipment, or perhaps computers that operate via the subtlest voice commands or body movements. They are reported here for sake of completeness but, for the moment, they are too premature to be proposed as real alternatives (Fig. 3.4).

Fig. 3.3 An example of EMG acquisition device. The advantage of using these signals over EEGs for the control of an exoskeleton is that input signals for the exoskeleton would be picked up directly from the motor units involved in the hand control; this is the reason why most of the techniques seen in literature use this approach to address the problem of robotic hands/exoskeletons control

Table 3.2 Pros and cons concerning the use of surface EMG	Pros	Cons
	Lot of research being done on this topic	Skin-electrode contact plays an important role
	The acquired signals are really specific to the user's hand actuation	

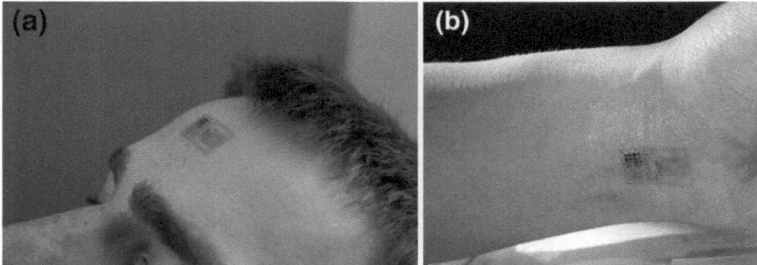

Fig. 3.4 a–b The "tattoo electronic" device developed by researchers of the Northwestern University and the University of Illinois. The device can be attached to the skin as a temporary tattoo to monitor brain, heart or muscle tissues, transferring data through wireless communication. "Hard" electronic is integrated into a flexible structure based on wave-like nano-ribbon constructions, to link the components by means of connections with a stiffness similar to the skin's one, able, therefore, to be flexed with no breakage. The devices can contain EEG, EMG or EKG sensors (reproduced with permission of J. Rogers, University of Illinois)

3.3 Alternative Designs

Several exoskeleton structure designs were found in literature; they solve some of the main problems but present a series of disadvantages concerning either size or feasibility of control.

3.3.1 Actuation

The *Cable mechanism* is based on the transmission of the movement by means of cables connected to the joint, or controlled part, and the motor. The force results in the movement of the cables, which leads to the joint rotation through a proper coupling. By means of cable transmission it is possible to avoid the presence of bulky structures on the hand, displacing the actuators in a more comfortable position, such as the forearm. However, some problems arise from the fact that cables work only in traction, so that two cables are needed to grant two different directions of movement (flexion and extension).

Orthotic devices have been designed to help patients with impaired motion functionalities in the most useful hand movement, pinching. A pinching device for the index finger is shown in the following figure. The exoskeleton is composed of three aluminum bands, placed in correspondence to the three finger's bones, and a plate mounted on the hand's dorsal surface. The three bands are movable, so that they can be adjusted to the user's hand size. The device consents operating the three joints, allowing flexion and extension; in particular, the distal and proximal interphalangeal joints (DIP, PIP) movements are coupled, while the metacarpo-phalangeal (MCP) joint is independent. Abduction and adduction are not mechanically controlled, but the movement is allowed, enabling the natural movement of flexo-extension. Motion is transmitted from the actuator (i.e. a pneumatic device, not suitable for our application) through steel cables running along the finger front and backside and pulled by a piston acting in compression. The MCP flexion is controlled by a floating linkage to the base plate and to a second piston, acting in extension and responsible for the joint extension move-ment (Lenny Lucas DiCicco and Matsuoka 2004) (Fig. 3.5).

The next figure shows a CAD drawing of the mechanical structure of one finger used in the rehabilitation hand exoskeleton developed by Wege et al. The system is based on a leverage system connected to the hand attachment on one side and to each phalanx of the finger attachment on the opposite side. The adjustment of the base point of the levers to the hand's size is performed by regulating a screw. The device controls the movements of flexion and extension of the MCP, PIP and DIP joints independently, and adduction and abduction of the MCP joint. The actuation is based on the use of 4 actuators (one for each degree of freedom) controlled by a DC motor operating in torque control mode. Force is transmitted by Bowden cables attached to a pulley moved by the motor through transmission gears; two

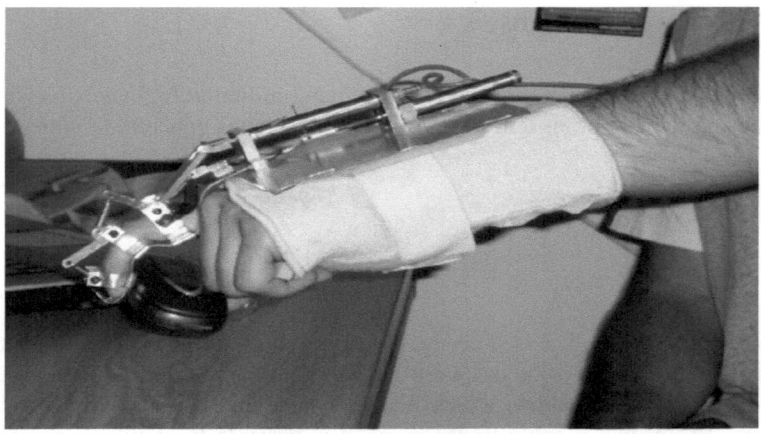

Fig. 3.5 The EMG-controlled hand exoskeleton for natural pinching developed by Lucas et al. The device is composed of three aluminum plates in correspondence to the finger bones and a base plate on the back of the hand. Flexion and extension movements of the three joints are allowed, and the motion is transferred from a pneumatic piston through steel cables and linkages (Lenny Lucas DiCicco and Matsuoka 2004)

Bowden sheaths for each joint are connected to a device that keeps the cables under tension. Two ends of the Bowden cable are fixed to a pulley of the levers and their movement within the sheaths leads to a rotation of the pulleys controlling the joints rotation, so that the motion of each phalange is based on the previous one, resulting in the affection of the later joint control by the preceding ones (Wege and Günter 2005) (Fig. 3.6).

In order to accurately control the joint rotation and avoid any interference between the structure and the hand/fingers, a technique known as the *four bars mechanism* is also used. The mechanism is shown below; as it can be observed, the interference with the fingers during the flexion movement can be avoided thanks to a parallelepiped structure, which allows to move the rotation center in correspondence of the considered joint's one (Fang et al. 2009). Considering our application, the advantages include the absence of hurdles on the hands' palms; unfortunately, the observed structures are generally bulky and quite big sized (Fig. 3.7).

Another particularly attractive method, based on the same principle of the cables system, is the tendon system, which can reproduce the human flexor-extension mechanism which connects muscles to joints by means of tendons and ligaments. The adoption of this system consents to extremely contain the dimensions and bulk of the device, as showed in Fig. 3.3, depicting the SEM^{TM}-Glove. Founded by a researcher from the Karolinska Institute and the Royal Institute of Technology in Stockholm in 2006, Bioservo developed the grip strengthening glove SEM^{TM}Glove (Soft Extra Muscle Glove) winning the Robotdalen Innovation Award 2012. The device is intended to be used by

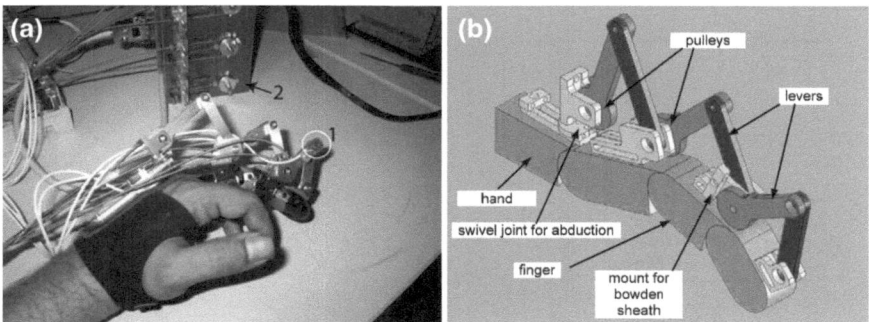

Fig. 3.6 Picture (**a**) and CAD drawing (**b**) of the hand exoskeleton for rehabilitation designed by Wege and Günter. The system is based on a leverage controlled by a DC motor in which each lever is controlled by a Bowden cable connected to a pulley controlling the rotation of each joint (Wege and Günter 2005)

individuals with impaired grip strength, which may be caused for example by muscles weakening, injured nerves or muscular pain. The functioning is based on force sensors placed on the fingertips, which record when the user is grasping an object and send the data to a microcontroller; this one calculates the needed force to be added and controls the tendons flexion by means of small size engines. The system is powered by two lithium–ion batteries. The control unit can be placed on the forearm or worn on the back (Bioservo 2006). The fingers flexion is enforced by the pulling effect exerted by tendons on the fingertips, as showed in Fig. 3.8.

In 2011 a joint project of NASA and General Motors led to the development of the first prototype of the Human Grasp Assist Device, or K-glove, designed to help astronauts in EVA by more than halving the needed pinching and grasping force, and thus reducing fatigue problems. The researchers began working within the Robonaut 2 project (R2), with the aim to provide the robot with the possibility to operate and handle tools designed for human use. This led to the employment of sensors, actuators and tendons comparable to the human muscles, nerves and tendons and offered a platform to the K-glove studies as well. The actuation system itself is inspired to the R2 one; actuators are placed in the upper portion of the gloves, while pressure sensors are located in correspondence to the fingertips (Dean and Flores 2012). A picture of the prototype is showed below (Fig. 3.9).

3.3.2 Feedback

In 2012, Med Sensation developed the Tricorder glove, based on the combination of several sensing systems and mainly aimed at medical diagnostic applications and students training. The glove combines a series of sensors, including accelerometers, force, temperature, sound and vibration sensors, and a data protocol for

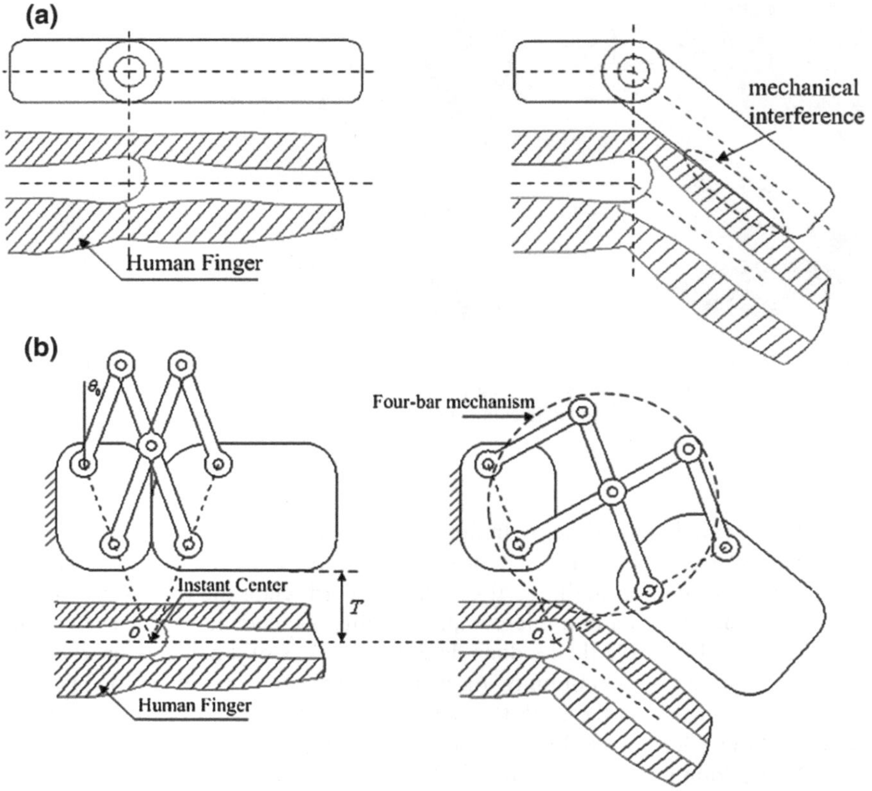

(a)

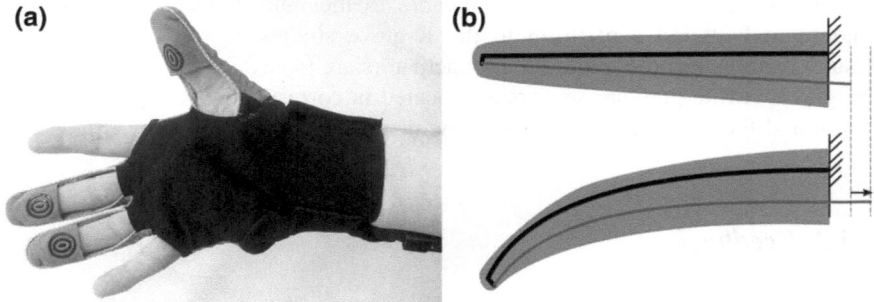

Human Finger

mechanical interference

(b)

θ_0

Four-bar mechanism

Instant Center

T

Human Finger

Fig. 3.7 Schematic of a four bars structure. The parallelepiped structure placed in correspondence to the joints allows to match the structure's rotation center to the joint's one, avoiding any interference with the finger. To obtain this matching the device has to be customized to the user's anatomy (Fang et al. 2009). **a** Parallel joint. **b** Four-bar mechanism joint

(a)   **(b)**

Fig. 3.8 **a** Photography of the SEM^{TM}Glove. **b** schematic of the tendon mechanism adopted; the force is exerted by pulling the tendons, which, being linked to the fingertips of the glove, lead to the flexion of the structure (reproduced with permission of Bioservo Technologies)

Fig. 3.9 The picture shows a prototype of the Human Grasping Assistance device, or K-glove, developed by a joint project of NASA and General Motors as a tool to assist astronauts in EVAs by decreasing the needed force for hand's movement in order to reduce the fatigue (Damir 2012) (NASA Courtesy)

Fig. 3.10 The image shows the second prototype of the Glove Tricorder developed by Med Sensation. Several sensors are located on the fingertips and hand palms, signals are elaborated in the unit placed on the forearm and transferred through a wireless protocol

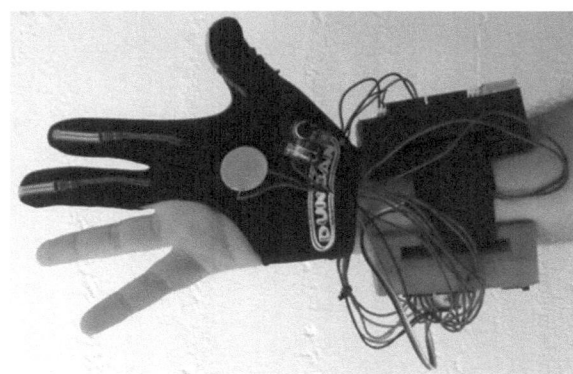

the wireless transmission of data. Students in medicine would take advantage in their practice by the quantification of touch, while patients would be given the ability to carry on self-diagnosis, assessing sports injury and performing self-clinical breasts exams. The team aims at a future development based on the addition of ultrasounds probes, able to integrate data allowing a real-time assessment of cardiac valves functioning or abdominal pain. A prototype is showed in Fig. 3.10, microphones, thermometer, accelerometers and ultrasound, flow, pressure and vibration sensors are placed on the fingertips, while a heat infrared camera is located on the hand palm (MedSensation 2012).

In 2012 the Ukrainian team QuadSquad won the Windows Image Cup competition developing a system able to translate signs language into text and spoken words. The device, called EnableTalk, is composed by two parts: the hardware part is constituted by a pair of gloves embedding 15 flex sensors, accelerometers, gyroscopes in order to determine the position of the hand in the space; the second is a data elaboration software for smartphones developed under Windows Phone 7/ Windows 8, and responsible to the conversion of gesture in words and sounds. The prototype include a Bluetooth module to transmit the date to a mobile device and a

(a)

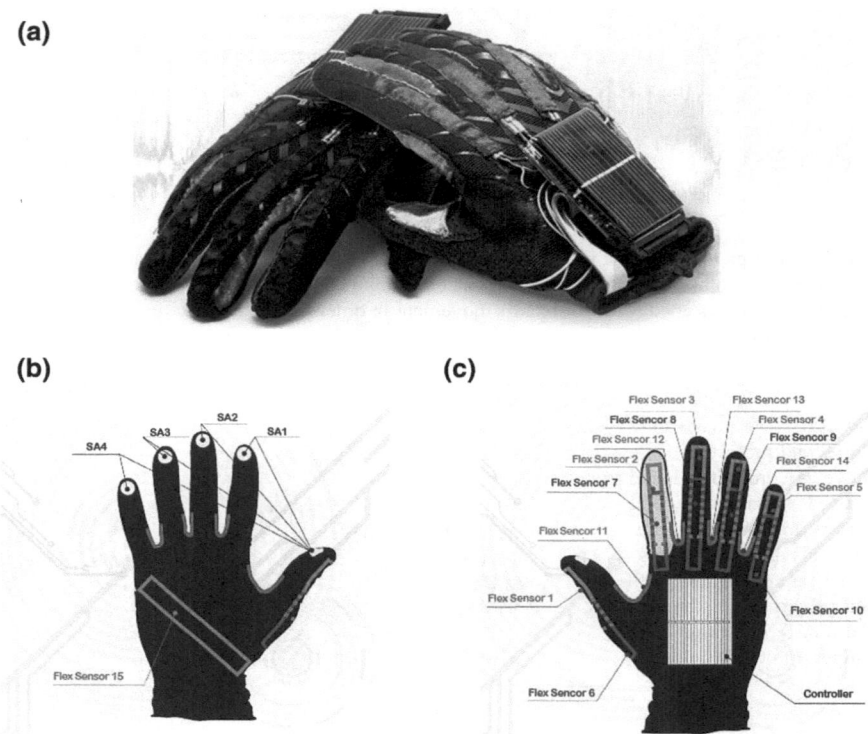

(b)   **(c)**

Fig. 3.11 a The EnableTalk prototype presented at Image Cup 2012. **b–c** The schematics illustrate the positioning of the embedded sensors within the glove (reproduced with permission of QuadSquad 2012)

USB port for the PC synchronization and to charge the Li–ion battery. QuadSquad estimates the price per device will drop to around \$20 if EnableTalk enters mass production (QuadSquad 2012) (Fig. 3.11).

References

Ball T, Kern M, Mutschler I, Aertsen A, Schulze-Bonhage A (2009) Signal quality of simultaneously recorded invasive and non-invasive EEG. NeuroImage, 46(3)
Bioservo (2006) http://www.bioservo.com. Accessed 29 August 2012
Boulenger V, Silber B, Roy A, Paulignan Y, Jeannerod M, Nazir T (2008) Subliminal display of action words interferes with motor planning: a combined EEG and kinematic study. J Physiol Paris 102:130–136
Damir B (2012) Human grasp assist device provides additional grip force. Tratto da http://www. robaid.com. Accessed 15 Mar 2012

Dean B, Flores D (2012) NASA, GM jointly developing robotic gloves for human use. Tratto da http://www.nasa.gov. Accessed 13 Mar 2012

Emotiv (n.d.) http://www.emotiv.com. Accessed 1 Jan 2012

Fang H, Xie Z, Liu H (2009) An exoskeleton for controlling DLR/HIT hand. Conference on intelligent robots and systems

Favetto A, Fai Chen Chen AE, Manfredi D, Calafiore G (2010) Towards a hand exoskeleton for a smart EVA glove. IEEE international conference on robotics and biomimetics (ROBIO)

Friesen G, Jannett T, Jadallah M, Yates S, Quint S, Nagle H (1990) A comparison of the noise sensitivity of nine QRS detection algorithms. IEEE Trans Biomed Eng 37:85–98

Henderson D (2007) Novel piezo motors enable positive displacement microfluidic pump. New Scale Technologies, US

Hochberg LR, Bacher D, Jarosiewicz B, Masse N, Simeral J, Vogel J et al (2012) Reach and grasp by people with tetraplegia using a neurally controlled robotic arm. Nature 485:372–375

Lenny Lucas DiCicco M, Matsuoka Y (2004) An EMG-controlled hand exoskeleton for natural pinching. J Robot Mech 16:482–488

MedSensation (2012) Tratto da http://www.medsensation.com

New Scale Technologies I (n.d.) SQUIGGLE micro motor technology. http://www.newscaletech.com. Accessed 2012

Northwestern (n.d.) http://www.nothwestern.edu. http://www.northwestern.edu/newscenter/stories/2011/08/tattoo-electronics-huang.html. Accesssed 20 Oct 2012

Porro C, Francescato M, Cettolo V, Diamond M, Baraldi P, Zuiani C et al (1996) Primary motor and sensory cortex activation during motor performance and motor imagery: a functional magnetic resonance imaging study. J Neurosci 16(23):7688–7698

QuadSquad (2012) http://enabletalk.com/abstract.html. Accessed 15 Sept 2012

Velliste M, Perel S, Spalding M, Whitford A, Schwartz A (2008) Cortical control of a prosthetic arm for self-feeding. Nature 453(7198):1098–1101

Vuskovic M, Li X (1996) Blind separation of surface EMG signals. Conference of the IEEE Engineering in Medicine and Biology Society

Wege A, Günter H (2005) Development and control of a hand exoskeleton for rehabilitation of hand injuries. Technische Universität Berlin, Berlin

Chapter 4
The Soft Robotics Approach

Abstract Starting from the user's requirements previously defined, a new soft robotics approach was chosen and developed in order to overcome the criticalities arisen in the analysis of the state of the art. One of the key points of soft robotics is biomimicry: in place of heavy, rigid and noisy motors, *artificial muscles* are in charge of the movement of the soft structure, allowing a number of degrees of freedom unthinkable with traditional mechanics. The shift from hard to soft robotics brings the focus on materials: actuating and sensing devices are embedded in the material itself, which turns out to be smart. In particular, the attention was focused on *Electroactive Polymers* (EAPs): these polymeric materials work as transducers, converting electrical inputs into mechanical outputs, and vice versa. After an extensive material selection procedure among all the EAPs solutions currently available, Dielectric Elastomers (DEs) emerged as the most suitable materials for the intended application and a mathematical model of their electro-mechanical behavior is presented. The control of the hand exoskeleton is addressed in this section. Its objective is to help the astronaut accomplish the tasks he has to perform. The entire control system is composed of four phases: the recognition of the astronaut's will (detect and distinguish the different movements), the control strategy, the enhancement of these movements and the measure of the actual position and force. The final measure system was implemented focusing on *redundancy*, *safety control* and to assure minimum performances also in off-nominal conditions.

The solution must not add complexity, rigidity or weight to the spacesuit glove. Therefore, a traditional *hard* approach is not suitable: a metal exoskeleton, operated with motors, pumps or compressors, would imply a significant increase in the mass of the spacesuit, as well as a higher stiffness of the system and a reduction of the astronaut dexterity. Thus, *soft robotics* appears as the best solution for this application. As the name suggests, systems built within this approach are soft, flexible and compliant. Furthermore, it is possible to abate the total weight of the

---

Chapter written with the contribution of Roberto Rossi (Politecnico of Milano).

P. Freni et al., *Innovative Hand Exoskeleton Design for Extravehicular Activities in Space*, PoliMI SpringerBriefs, DOI: 10.1007/978-3-319-03958-9_4, © The Author(s) 2014

exoskeleton and to get rid of rigid elements constraints. One of the key points of soft robotics is biomimicry: in place of heavy, rigid and noisy motors, *artificial muscles* are in charge of the movement of the soft structure, allowing a number of degrees of freedom unthinkable with traditional mechanics.

On the basis of information and data collected analyzing the state of the art and the users' requirements, a qualitative comparison between hard and soft robotics has been performed. Before embracing the soft approach, efforts and time were devoted to the problem setting and to the identification of the critical issues to be addressed. The result of this analysis is presented in Fig. 4.1.

The soft robotics approach has many advantages and, despite some important drawbacks, the overall trade-off is in its favor. One of the main goal pursued during the project development was to overcome some critical issues related to the soft robotics.

The hand movements to be supported and augmented concern the first three fingers: thumb, index finger and middle finger. In fact, these are the most used fingers during Extra-Vehicular Activities and, consequently, they are subject to fatigue occurrence. The main efforts in developing the solution were focused on the index finger and, in particular, on the hinge joints between its phalanges.

The control strategy architecture was designed to be seamless and redundant, in order to ensure that the device could be directly controlled by the astronaut's will and did not cause damage to the user's hand.

The development of the solutions was performed by following three main pathways:

(1) Materials for actuation and sensing.
(2) Measure system.
(3) Control system architecture.

In the following sections, each topic is analyzed and discussed in detail, providing a deep insight into the reasons that led to design decisions, along with the rationale behind them.

4.1 Materials for Actuation and Sensing

The shift from hard to soft robotics brings the focus on materials: actuating and sensing devices are embedded in the material itself, which turns out to be smart. *Smart materials* represent a wide class of materials, belonging to different families (polymers, metal alloys and ceramics) and characterized by a large number of control mechanisms. They can respond to external stimuli of diverse nature, like an applied stress or an imposed strain, a change in temperature, relative humidity or pH of the environment, a magnetic or electric field and so forth. In this variety of possibilities, materials with very different properties may be found.

In the framework of soft robotics, and taking into account the requirements of the project, polymeric smart materials appeared to be the most suitable choice for the

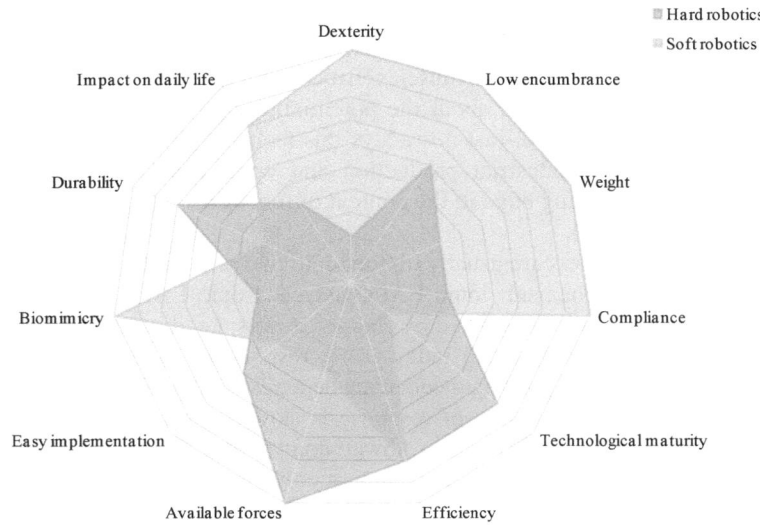

Fig. 4.1 Multi criteria comparison of soft and hard robotics

Fig. 4.2 Selection process

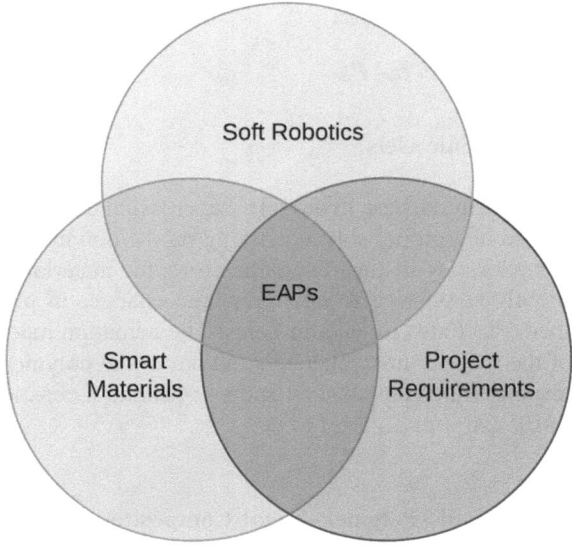

application under study (Fig. 4.2). In particular, the attention was focused on *Electroactive Polymers* (EAPs): these polymeric materials work as transducers, converting electrical inputs into mechanical outputs, and vice versa. They show features that cannot be traced in other *traditional* functional materials (e.g. piezo-electric ceramics), such as large active strains, high energy density, mechanical compliance and flexibility, very low weight, zero noise emission, simple and scalable structures and tailorable properties. All these characteristics make EAPs the actuation system closest to natural muscles (Brochu and Pei 2010; Carpi et al. 2011).

Despite the first awareness about the potentialities of Electroactive Polymers may be traced back to 1880, just nowadays this technology is moving from academic laboratories to industrial production and commercialization. Therefore, EAPs represent the cutting edge in the context of functional materials (Carpi et al. 2011; Bar-Cohen 2002).

Electroactive polymers are usually divided into two principal classes, according to their actuation mechanism: ionic EAPs and electronic EAPs. The first group relies on electric activation mediated by charge carriers, i.e. ions and/or molecules, while materials belonging to the latter group respond to the stimulation of an electric field. When the triggering mechanism depends on the diffusion of relatively bulky chemical species, the rate of response is slowed down. Consequently, the actuation speed of ionic EAPs is much slower, compared to electronic EAPs: this is their major drawback. On the other hand, smart polymers actuated by electric fields are able to apply relatively small forces and require very high voltages.

A brief look at some electroactive polymers may be useful to get a general picture about the technology and to analyze advantages and drawbacks of different alternatives.

4.1.1 Ionic EAPs

4.1.1.1 Ionic Gels

These materials are hydrogels, i.e. crosslinked polymers which are able to swell if placed in a proper solvent. Acting on the polymer–liquid interaction, the swelling behavior may be tuned and, therefore, the material can be actuated. Hydrogels are usually responsive to environmental changes in pH or temperature and some of them react also to electric fields. The actuation mechanism relies on the diffusion of the solvent molecules into and out of the polymer network and, for this reason, response rate is quite slow and encapsulation constitutes an issue (Brochu and Pei 2010; Carpi et al. 2011).

4.1.1.2 Ionic Polymer–Metal Composites

Ionic polymer–metal composites (IPMCs) are made up of a polymer ion exchange membrane, sandwiched between two compliant metal electrodes (percolated nanoparticle of Pt or Au). This technology is based on the peculiar properties of the membrane: it is characterized by an interconnected porosity which let solvents move through it. This way, when a bias voltage is applied, the ions migrate to the oppositely charged electrode, making the polymer swell on one side and shrink on the other, with a net bending of the material. The actuation needs very low tension (in the order of few volts), but the achievable strain is modest. Therefore their application as artificial muscles is limited (Brochu and Pei 2010; Carpi et al. 2011).

4.1.1.3 Conducting Polymers

Even if most polymeric materials are insulators, some of them have a chemical structure which turns them into conductors. In particular, the presence of alternated double bonds along the backbone chain (conjugated system) leads to the formation of a *macromolecular orbital*: the π-bonds give rise to a wide resonance structure and the delocalization of the electrons lets charge carriers move almost freely along the polymer chain. Polypyrrole and polyaniline (PANI) are the most widely used conductive polymers. Nevertheless, conductive polymers behave more like semiconductors than pure conductors and, furthermore, they can undergo a reversible doping through redox reactions. The actuation of this electroactive polymer is based precisely on electrochemical processes: by means of doping, electrical charges are localized throughout the chain and the uptake of counterions causes an increase in volume, perpendicularly to the chains orientation. This requires the presence of a liquid electrolyte that has to flow through the polymer. Conductive polymers suffer from very low operating efficiency (around 1 %), limited actuation speed and encapsulation issues (Brochu and Pei 2010; Carpi et al. 2011).

4.1.1.4 Carbon Nanotubes

Carbon nanotubes (CNTs) are nowadays used in a really wide range of applications, thanks to their peculiar features. In fact, they show outstanding mechanical properties (Young modulus up to 1 TPa) and high conductivity. Moreover, they may be functionalized and used as transducer: if suspended in an electrolyte, carbon nanotubes are able to enlarge thanks to double-layer charge injection. The mechanism is activated applying a potential between CNTs and a counterelectrode; the polarization of CNTs surfaces generates an ion migration in the electrolyte in order to recover the overall electrical neutrality of the system. Due to their high stiffness, achievable strains are small (<2 %); on the other hand, required voltage are low as well (≈ 1 V) (Brochu and Pei 2010; Carpi et al. 2011).

4.1.2 Electronic EAPs

4.1.2.1 Piezoelectric Polymers

The molecular structure of piezoelectric polymers contains permanent dipoles, due to the presence of high electronegative halogens atoms in the side groups. These dipoles can be aligned by means of an electric field and, as a result, the material can be polarized permanently. Furthermore, the polarization can be cancelled out applying an opposite bias or bringing the material above a defined temperature, called Curie temperature. In particular, if the material is above its Curie point, a reversible transition between polar and non-polar phase is possible. This is the property exploited to use those materials as actuators. Polyvinylidene difluoride (PVDF) is one of the most common piezoelectric polymers: above its Curie point, which is close to room temperature, the transition between the β paraelectric phase (non-polar) and the α ferroelectric phase (polar) can be induced with an electric field. The phase transition is accompanied by a change in the morphological structure of the polymeric chain and eventually leads to a volume variation. PVDF has lower piezoelectric performances as compared to the ceramic counterpart, but, as a polymer, it is more lightweight and flexible, easier to form and capable to produce two orders of magnitude larger strains. One of the major concerns regarding these materials is the great energy loss due to hysteresis phenomena during polarization (Brochu and Pei 2010).

4.1.2.2 Electrostrictive Polymers

The materials belonging to this group are copolymers based on vinylidene fluoride (VDF) and trifluoroethylene (TrFE) with the same actuation mechanism of piezoelectric polymers. P(VDF-TrFE) has the advantages of a chemically tailorable Curie temperature and a lower energy barrier for the α–β phase transition. This results in reduced hysteresis energy loss, higher efficiency and more flexibility on operating temperatures (Brochu and Pei 2010).

4.1.2.3 Liquid Crystal Elastomers

These polymers are characterized by the presence of mesogen phases (typical of liquid crystals) within an elastomeric network. Mesogens are rigid segments that can be oriented in partly ordered structures via thermal or electrical stimuli, causing a deformation of the material. Thermally activated LCEs show very high strains (up to 400 %), but their effectiveness is limited by the need for heat diffusion, which slows down the response rate. On the other hand, electrically activated LCEs are much faster, but they can reach lower strains (Brochu and Pei 2010; Carpi et al. 2011).

4.1.2.4 Dielectric Elastomers

The working principle underlying dielectric elastomers is the same of a capacitor. A dielectric elastomeric thin sheet is covered on both sides with conductive flexible material in order to create a capacitor characterized by very compliant electrodes. When a voltage is applied between the two electrodes, the opposite charges on the two faces are attracted, while the like charges on each side repel each other. The resulting effect on the elastomer is a net contraction in thickness and a planar expansion. The actuation voltages are very high (in the order of kilovolts), but the capacitive nature of the solution guarantees little currents and, therefore, very low power consumption. The output stress increases quadratically with the applied electric field and large strains may be achieved. Both silicone rubbers and acrylic elastomers are used as dielectric elastomers. Furthermore, applying a prestrain to the material, efficiency can be highly improved and preferential actuation in one direction can be realized (Brochu and Pei 2010).

4.1.2.5 Carbon Nanotube Aerogels

Differently from the ionic carbon nanotubes, the aerogel solution does not require a liquid electrolyte: carbon nanotubes are parallel arranged in ordered *forests*. Applying a positive voltage between the CNTs and a grounded electrode, the like charges generated on the surface of nanotubes repel each other causing a transversal expansion. This material has an incredibly low density and it is capable of high and quick strains. One of the few drawbacks is represented by the relatively low forces generated (Aliev et al. 2009; Carpi et al. 2011).

4.1.3 Materials Selection

The literature about electroactive polymers is still fairly poor in numerical data because these materials are quite recent and the research in the field is fragmented. Therefore, the considered properties have been measured in different conditions and can be used just for a qualitative comparison.

One of the most important properties for the considered application is the *maximum strain*. In fact, artificial muscles shall be flexible and able to undergo large deformation without getting damaged. The maximum strain is the largest displacement reached, normalized to the initial length of the sample in the direction of actuation. Together with the strain, also the stress that the material is able to sustain is important. As far as actuators are concerned, the *maximum pressure* is the largest force per cross-sectional area that the material is capable to develop. Usually, the peak strain and the peak stress do not correspond: it is common to measure the greatest force at zero strain and to have a null stress when the deformation is at its maximum.

To compare different actuation solutions, another metrics is more appropriate: the *actuator energy density*, i.e. "the maximum mechanical energy output per cycle and per unit volume of material" (Pelrine et al. 2000a, b). This feature takes into account both stress and strain, giving a more comprehensive description of the material. Nevertheless, the *elastic energy density*, equal to the half of the actuator energy density, is conventionally used. Moreover, if the energy is referred to unit mass (dividing it by density), the *Specific elastic energy density* is obtained.

The *Coupling efficiency*, k^2, is defined as the ratio between the energy converted into mechanical work per cycle and the electrical energy provided per cycle. It is useful to guarantee a consistent comparison with traditional piezoelectric materials. Low values of this parameter imply that high efficiency in actuation cannot be reached due to large energy loss (Pelrine et al. 2000a, b).

For the artificial muscles application, the actuation speed of the material is not negligible. In fact, fast response is needed to guarantee performances as close as possible to natural muscles and to develop a device reliable in a wide range of movements.

An overview of data available in the literature is presented in Table 4.1. Natural muscles are used as a reference and other technologies are reported for comparison.

At a first sight, some of the electronic EAPs show the largest strains and, even though they do not register the best performances in the maximum pressure, their mechanical properties are comparable with (and slightly superior to) those of natural muscles. For a more comprehensive analysis and coherent selection a multicriteria approach was adopted. The considered properties were: maximum strain, maximum pressure, elastic energy density and relative speed. Furthermore, the values of maximum pressure were normalized on the density, in order to have more homogeneous data.

Data span on such a wide range of values that, to perform an effective comparison, their logarithmic values (on basis 10) needed to be used for the subsequent considerations. Each property of the materials was ranked with a grade, in a scale from 1 to 10, covering the whole range of values so that the lowest one scored 1 and the highest one scored 10. Afterwards, a weighted average was computed and a final grade was assigned to each material (Table 4.2). A weighting factor of 2 was assigned to the maximum strain because flexibility and device compliance are of major relevance for the considered application. All the other properties had the same weight (weighting factor of 1). Due to lack of data, some options were penalized, but the results are acceptable for a qualitative assessment.

In this evaluation the *Natural muscle* (*peaks in nature*) scores 6/10: therefore, all the alternatives with higher grades are "sufficient" and suitable candidates for artificial muscles applications, even though further considerations are needed.

The ranking list is reported in Table 4.3. The top performing option is represented by *Carbon Nanotubes Aerogels*. This is a promising technology, even though at the very first stage of development: it gets the maximum score (10) in pressure provided, strain and actuation speed. Data about energy are missing and further investigations are needed. Thanks to its dramatically low density due to the aerogel formulation, stiffness provided by carbon nanotubes and large achievable strains enabled by the actuation mechanism, CNT Aerogels are potentially the best option. The state of the art of the technology is still too immature for an application in the short-medium term and, therefore, it was disregarded for the moment.

Table 4.1 Electroactive polymers: overview on data available in literature

	Type (specific)	Maximum strain %	Maximum pressure, Mpa	Specific elastic energy density J/g	Elastic energy density J/cm³	Coupling efficiency k² %	Maximum efficiency %	Specific density g/cm³	Relative speed (full cycle)	References
Natural reference	Natural muscle (human skeletal)	40	0.35	0.07	0.07	–	35	1	Medium	Pelrine et al. (2000a, b)
	Natural muscle (peaks in nature)	100	0.8	0.04	0.04	–	40	1	Slow–fast	Madden et al. (2004), Brochu and Pei (2010)
Ionic EAPs	Ionic gels (polyelectrolyte)	40	0.3	0.06	0.06	–	30	1	Slow	Pelrine et al. (2000a, b)
	IPMC	3	30	–	–	–	–	1	–	Brochu and Pei (2010)
	Conducting polymer (PANI)	10	450	23	23	1	5	1	Slow	Pelrine et al. (2000a, b), Brochu and Pei (2010)
	Carbon nanotubes	2	26	200	200	–	–	1	Fast	Brochu and Pei (2010), Carpi et al. (2011)
Electronic EAPs	Piezoelectric polymer (PVDF)	10	4.8	0.0013	0.0024	7	–	1.8	Fast	Pelrine et al. (2000a, b)
	Electrostrictive polymer [P(VDF–TrFE)]	4.3	43	0.49	0.92	–	80	1.8	Fast	Pelrine et al. (2000a, b)
	Liquid crystal elastomers	4	–	–	0.02	–	–	–	Fast	Brochu and Pei (2010)
	Dielectric elastomer (acrylic with prestrain)	380	7.2	3.4	3.4	85	60–80	1	Medium	Madden et al. (2004), Pelrine et al. (2000a, b), Brochu and Pei (2010)
	Dielectric elastomer (silicone with prestrain)	63	3	0.75	0.75	63	90	1	Fast	Brochu and Pei (2010)

(continued)

Table 4.1 (continued)

Type (specific)	Maximum strain %	Maximum pressure, Mpa	Specific elastic energy density J/g	Elastic energy density J/cm³	Coupling efficiency k² %	Maximum efficiency %	Specific density g/cm³	Relative speed (full cycle)	References
Dielectric elastomer (silicone-nominal prestrain)	32	1.36	0.22	0.2	54	90	1	Fast	Pelrine et al. (2000a, b)
Carbon nanotubes aerogels	220	16	0.03	–	–	–	0.0015	Fast	Aliev et al. (2009)
Piezoelectric ceramic (PZT)	0.2	110	0.013	0.1	52	90	7.7	Fast	Pelrine et al. (2000a, b)
Piezoelectric single crystal (PZT-PT)	1.7	131	0.13	1	81	90	7.7	Fast	Pelrine et al. (2000a, b)
Shape memory alloy (TiNi)	5	200	15	100	5	10	6.5	Slow	Pelrine et al. (2000a, b)
Magnetostrictive (Terfenol-D)	0.2	70	0.0027	0.025	–	60	9	Fast	Pelrine et al. (2000a, b)
Electrostatic devices (integrated force array)	50	0.03	0.0015	0.0025	50	90	1	Fast	Pelrine et al. (2000a, b), Brochu and Pei (2010)
Electromagnetic (voice coil)	50	0.1	0.003	0.025	–	90	8	Fast	Pelrine et al. (2000a, b), Brochu and Pei (2010)

Other technologies

Table 4.2 Grades for each material properties

	Type (specific)	Maximum pressure/ρ	Maximum strain	Elastic energy density	Relative speed (full cycle)	Final grade
Natural reference	Natural muscle (human skeletal)	3	8	3	5	5.40
	Natural muscle (peaks in nature)	4	9	3	5	6.00
Ionic EAPs	Ionic gels (polyelectrolyte)	3	8	3	1	4.60
	IPMC	6	4	0	0	2.80
	Conducting polymer (PANI)	8	6	9	1	6.00
	Carbon Nanotubes	6	4	10	10	6.80
Electronic EAPs	Piezoelectric polymer (PVDF)	4	6	1	10	5.40
	Electrostrictive polymer [P(VDF–TrFE)]	6	5	6	10	6.40
	Liquid crystal elastomers	0	4	2	10	4.00
	Dielectric elastomer (acrylic with prestrain)	5	10	7	5	7.40
	Dielectric elastomer (silicone with prestrain)	5	8	6	10	7.40
	Dielectric elastomer (silicone-nominal prestrain)	4	7	4	10	6.40
	Carbon Nanotubes Aerogels	10	10	0	10	8.00
Other technologies	Piezoelectric ceramic (PZT)	6	1	4	10	4.40
	Piezoelectric single crystal (PZT-PT)	6	3	6	10	5.60
	Shape memory alloy (TiNi)	6	5	10	1	5.40
	Magnetostrictive (Terfenol-D)	5	1	3	10	4.00
	Electrostatic devices (integrated force array)	1	8	1	10	5.60
	Electromagnetic (voice coil)	1	8	3	10	6.00

Table 4.3 Ranking list of the alternative materials

Rank	Type (specific)	Final grade	Rank	Type (specific)	Final grade
1	Carbon nanotubes aerogels	8	11	Electrostatic devices (integrated force array)	5.6
2	Dielectric elastomer (silicone with prestrain)	7.4	12	Shape memory alloy (TiNi)	5.4
3	Dielectric elastomer (acrylic with prestrain)	7.4	13	Piezoelectric polymer (PVDF)	5.4
4	Carbon nanotubes	6.8	14	Natural muscle (human skeletal)	5.4
5	Electrostrictive polymer [P(VDF–TrFE)]	6.4	15	Ionic gels (polyelectrolyte)	4.6
6	Dielectric elastomer (silicone-nominal prestrain)	6.4	16	Piezoelectric ceramic (PZT)	4.4
7	Conducting polymer (PANI)	6	17	Magnetostrictive (Terfenol-D)	4
8	Electromagnetic (voice coil)	6	18	Liquid crystal elastomers	4
9	Natural muscle (peaks in nature)	6	19	IPMC	2.8
10	Piezoelectric single crystal (PZT-PT)	5.6			

The second and the third places are occupied by *Dielectric elastomers*. Both solutions require a prestrain of the material to guarantee high performances and, in this respect, the possibility to tailor a preferential direction of actuation represents a plus, according to the project requirements. Furthermore this technology is already available on the market and its development has already reached a quite mature state.

Carbon nanotubes have such a high grade mainly thanks to the great elastic energy density and the fast actuation response. Nevertheless, they are not appropriate for the project application because they need a liquid electrolytic medium to work, with all the resulting complications (encapsulation of the device), and they have a very poor maximum strain.

Electrostrictive polymer [*P(VDF-TrFE)*] scores a sufficient mark, thanks to the relatively high pressure that it is able to provide and the fast response. Anyway, it still has some drawbacks: the maximum strain achievable, reported in literature, is small and the actuation mechanism is highly temperature dependent. The latter feature represents an important constraint for space application because proper working cannot be guaranteed in presence of large thermal excursions, typically registered during extra-vehicular activities.

PANI, a conductive polymer, results sufficient: it is characterized by the highest maximum pressure and it shows good elastic energy density, but its actuation response is slow and its maximum strain limited.

IPMC and *Liquid Crystals Elastomers* pay the price of missing information, resulting in an underestimation of their actual performance. For this reason, further investigations and, possibly, testing are required.

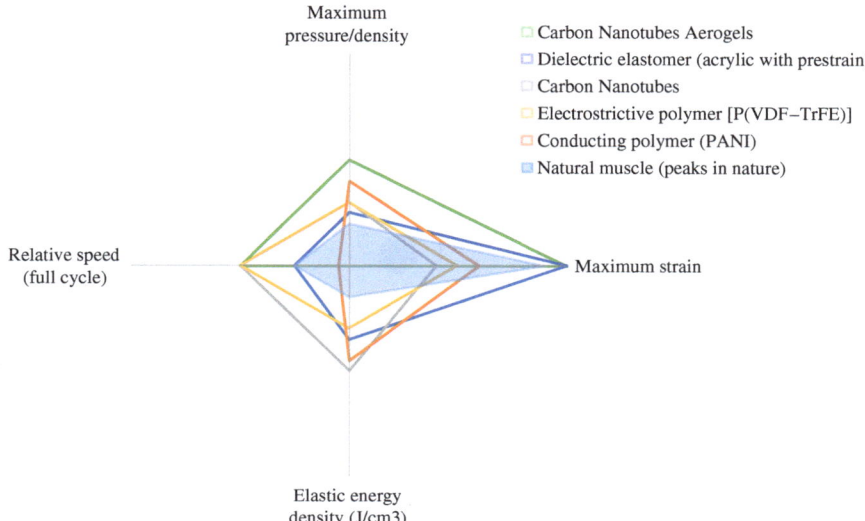

Fig. 4.3 Radar chart of the top-score options

In order to have a graphical outlook, the top-score results have been reported in Fig. 4.3.

Each shape represents a material and its dimensions are given by the score obtained in the four considered properties: the bigger is the shape, the better is the performance of each alternative. The reference, *Natural muscle (peaks in nature)*, is indicated by the filled quadrangle. Despite the *Carbon Nanotubes Aerogel* presents the largest area, the *Dielectric elastomer (acrylic with prestrain)* has a shape analogous to the natural muscle's one, but larger: this is an additional confirmation of the fact that the Dielectric Elastomers option is consistent with the reference and, therefore, adequate for artificial muscles application.

Further considerations for the material comparison may be drawn from a Pareto efficiency[1] analysis of the available alternatives. For each material, two parameters (to be optimized) were chosen: maximum pressure, divided by density, and maximum strain. The inverse of the logarithm on base 10 of both quantities was then computed and taken into account. The resulting plot is reported in Fig. 4.4: each bubble represents a candidate material and the alternatives which perform better, with respect to the parameters considered, are the closest to the axes origin. Looking at the chart, there is no option that minimizes, at the same time, both the parameters; therefore, a trade-off curve (Pareto front) was traced. The bubbles on

---

[1] In the engineering context, Pareto efficiency is a useful selection method. Each option is first assessed under multiple criteria and then a subset of options is identified with the property that no other option can categorically outperform any of its members.

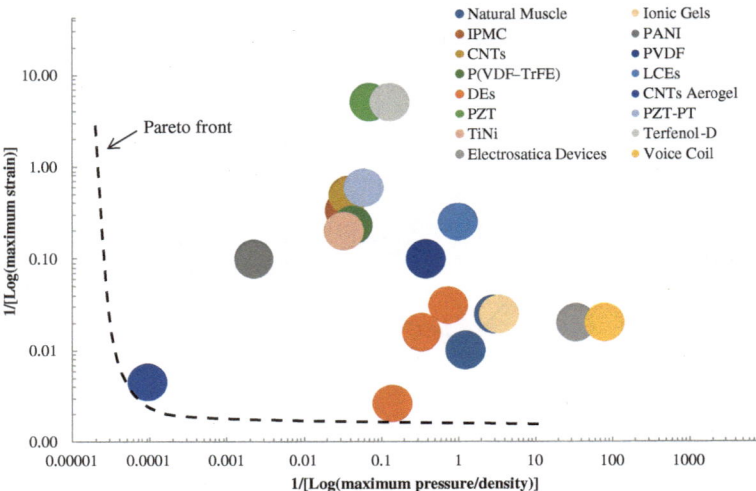

Fig. 4.4 Materials bubble chart with Pareto front

the Pareto front are *not dominated* by any other possibility and they are Pareto efficient, while all the others are *dominated* and do not optimize the performance.

The non dominated solutions are the *Carbon Nanotubes Aerogel* and the *Dielectric elastomer (acrylic with prestrain)*. These materials present the best combinations of maximum pressure and maximum strain and, in this context, are the best alternatives.

According to the available data and the previous considerations, the *dielectric elastomers* represent the best compromise for the intended application. In fact, they allow large strains and both maximum pressure and energy density are superior to (but of the same order of magnitude of) the natural muscles' ones. Those features make them the favorite candidates to perform as artificial muscles, intended to help astronauts in their extra-vehicular activities: in fact they should provide enough pressure without further limiting dexterity and with no risk of hurting the user.

4.1.3.1 Dielectric Elastomers Materials

Elastomeric materials are usually characterized by a hyperelastic behavior, i.e. the stress–strain curve presents three regions (Fig. 4.5): a first linear region with a relatively "high" stiffness, followed by a *plateau* region, in which the Young modulus drops, and culminating in a *stretch-hardening* region, in which the elastomer becomes stiffer again.

Among the candidates for dielectric elastomer actuators, three groups of polymers were identified: silicones, polyurethanes and acrylic elastomers. The mechanical and electrical properties of few representatives of those groups of materials are presented in Table 4.4.

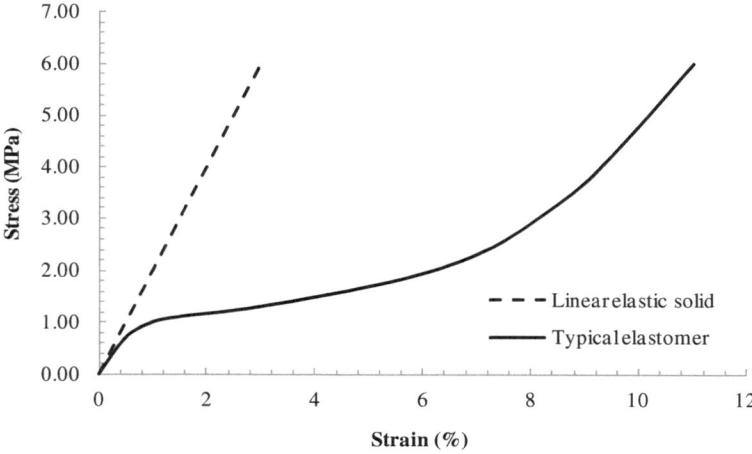

Fig. 4.5 Typical stress–strain curve for an elastomer

4.1.3.2 Effects of Prestrain on Dielectric Elastomers

Prestraining the elastomeric dielectric film, it is possible to increase its properties and improve the functioning of the actuator device as a whole. The material benefits from the prestrain application in terms of higher breakdown strength, better mechanical efficiency and faster actuation response, with just a limited reduction of the dielectric constant (Brochu and Pei 2010).

Furthermore, by means of prestrain, a tailorable anisotropy can be introduced in the material. With a proper design of the device, it is therefore possible to reach a preferential direction of actuation "by applying high prestrain in the direction perpendicular to the desired actuation direction and low prestrain along the actuation direction" (Brochu and Pei 2010).

Although prestrain brings many advantages, it requires a framing structure to keep the material in mechanical tension during actuation. This adds complexity to the device itself, increasing its mass and volume, and needs a focused design process.

4.1.3.3 Space-Related Issues

Polymers are the most critical materials as far as outgassing and atomic oxygen issues are concerned. They absorb humidity, depending on their chemical structure. They are particularly subject to radiation degradation: some chemical groups present in polymeric chains absorb UV radiations and undergo photochemical reactions, chain divisions or reticulation. These effects lead to changes in mechanical and physical properties of the material, in function of the total absorbed dose. Therefore, it is strongly recommended that any sensitive device is protected by a shield, that in the case of EVAs is the astronauts' suit.

Table 4.4 Comparison of some dielectric elastomers properties

Polymer (specific type)	Prestrain (x %, y %)	Energy density MJ/m³	Maximum actuation pressure Mpa	Thickness strain %	Area strain %	Young's modulus MPa	Breakdown electric field MV/m	Dielectric constant	Efficiency %	References
Silicone (Nusil CF19-2186)	(45, 45)	0.75	3	39	64	1.0	350	2.8	79	Pelrine et al. (2000a, b)
Polyurethane (Deerfield PT6100S)	–	0.087	1.6	11	–	17	160	7	–	Pelrine et al. (2000a, b)
Acrylic (3 M VHB 4910)	(300, 300)	3.4	7.2	61	158	3.0	412	4.8	80	Pelrine et al. (2000a, b)

4.1.4 Dielectric Elastomers Modeling

As previously mentioned in Sect. 5.1.2.4, the working principle of a dielectric elastomers (DEs) actuator is grounded in its capacitive nature: it is modeled as a parallel plate capacitor with dielectric made of a linearly elastic material. The application of bias causes a contraction in thickness of the dielectric, due to the electrostatic attractive force between the two electrodes, and a planar expansion in the other two directions (Fig. 4.6).

The dielectric material can be reasonably assumed to be incompressible, therefore:

$$l \cdot w \cdot t = cost \tag{4.1}$$

where l is the length, w the width and t the thickness of the dielectric. Or alternatively:

$$(1 + s_x)(1 + s_y)(1 + s_z) = 1 \tag{4.2}$$

where s_x, s_y and s_z are the strains in the three directions.

The electrostatic stress generated across the electrodes is the Maxwell pressure, described by:

$$p = \varepsilon_r \cdot \varepsilon_0 \cdot E^2 \tag{4.3}$$

where ε_r is the material relative dielectric constant, ε_0 is the vacuum dielectric constant and E is the applied electric field. Furthermore, the previous equation can be written as a function of voltage V and dielectric thickness t:

$$p = \varepsilon_r \cdot \varepsilon_0 \cdot \left(\frac{V}{t}\right)^2 \tag{4.4}$$

4.1.4.1 Free Boundary Condition

To a first approximation, the dielectric material was considered linearly elastic and isotropic (with a Young modulus Y) and a free boundary condition[2] was assumed for the actuator. Under those hypotheses, the thickness strain is given by:

$$s_z = -\frac{p}{Y} = -\frac{\varepsilon_r \cdot \varepsilon_0 \cdot E^2}{Y} = -\frac{\varepsilon_r \cdot \varepsilon_0 \cdot (V/t)^2}{Y} \tag{4.5}$$

---

[2] When no constraint is applied to the DE actuator, the electrostatic pressure (Eq. 4.3) strains the polymeric film until the elastomer's elastic stress (spring back effect) prevents further expansion.

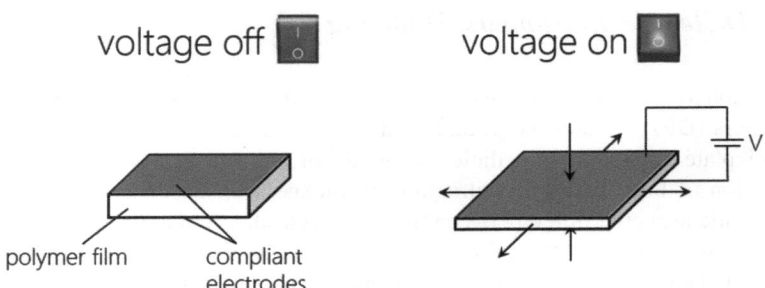

Fig. 4.6 Working principle of dielectric elastomers

This simple model is valid only for small strains (<10 %). For larger deformation, maintaining the linear elasticity assumption, another model has been put forward by (Pelrine et al. 1998; Brochu and Pei 2010):

$$s_z = \frac{2}{3} + \frac{1}{3}\left[f(s_{z0}) + \frac{1}{f(s_{z0})}\right] \tag{4.6}$$

where

$$f(s_{z0}) = \left[2 + 27s_{z0} + \frac{\left[-4 + (2 + 27s_{z0})^2\right]^{1/2}}{2}\right]^{1/3} \tag{4.7}$$

and

$$s_{z0} = -\frac{p}{Y} = -\frac{\varepsilon_r \cdot \varepsilon_0 \cdot E^2}{Y} = -\frac{\varepsilon_r \cdot \varepsilon_0 \cdot (V/z_0)^2}{Y} \tag{4.8}$$

In the free boundary condition the strains in the x and y direction can be considered equal

$$s_x = s_y = s_a \tag{4.9}$$

and, therefore, recalling the constant volume constraint (Eq. (4.2)) and solving for s_a:

$$s_a = (1 + s_z)^{-0.5} - 1 \tag{4.10}$$

A comparison of the two models was performed (Fig. 4.7): the large deformation Eq. (4.6) provided values of strain higher than those resulting from the equation valid for small deformations (4.5).

Both the previously described models are Hookean, i.e. they assume the Young modulus to be constant throughout all the deformation process. Actually,

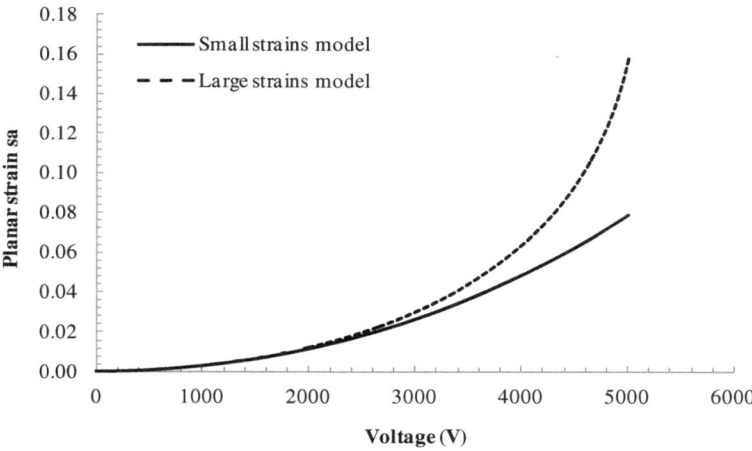

Fig. 4.7 Small strains and large strains models compared

elastomeric materials usually show a viscoelastic, time-dependent behavior and are non-Hookean. More accurate models, which take into account those features of the material, have been developed (Brochu and Pei 2010), but for the preliminary assessment aimed in this work, the Hookean model represents a reasonably good approximation. Furthermore, according to their typical mechanical behavior, elastomers show a decrease in elastic modulus with increasing strains. Therefore, the application of a Hookean model gives an overestimation of the material stiffness and the predicted strains will be lower than those really achievable during actuation (Eq. (4.5) shows that strain and Young modulus are inversely proportional). Eventually, the use of the simple Hookean approach will bring to an underestimation of the actuation strains and to an over dimensioning of the system (conservative approach).

4.2 Measure System

The control of the hand exoskeleton is addressed in this section. Its objective is to help the astronaut accomplish the tasks he has to perform. The entire control system is composed of four phases: the recognition of the astronaut's will (detect and distinguish the different movements), the control strategy, the enhancement of these movements and the measure of the actual position and force.

The measure systems are involved in the first and last phases, acting as inputs of the control and as tools for safety and performance control. The involved measures are obviously position and force, but also the uncommon will and state of fatigue, to be better specified. In fact, in the first instance, it is necessary to recognize which hand movements the astronaut *wants* to perform, the amount of force willed to apply and whether muscles are giving any signals of fatigue. The measure of the will and fatigue of the astronaut is a complex task, but literature and the study of body signals help solve this problem.

Summing up, the objectives of the entire measure system are:

- Detecting the will of a movement.
- Correctly identifying the movement comparing it with known previously recorded features.
- Measuring its intensity.
- Measuring the state of fatigue of muscles.
- Measuring the actual force and position, in order to assure safety and to control that the overall control strategy is well performing.

Descending from the task to be performed and the operational conditions, general requirements for the measure system are straightforward. Without reminding all the conditions identified in the EVAs, two critical conditions dependent from the hard physical activity of an astronaut have to be underlined: the presence of water and the change of human skin characteristic with fatigue and sweat.

The first characteristic required for a measure system is *repeatability*. This is highly influenced by the reproducibility of measure process with respect to the fatigue and sweat factors. *Precision* and *accuracy* are a must, because of the high importance and high risk tasks. *Low consumption, low weight* and *small dimensions* are required as for any component of the hand exoskeleton. *Redundancy* is needed to assure the safety of human beings and good performance in conditions different form the nominal ones.

The highest difficulty tasks to be performed are the measure of will and fatigue. Fortunately, body signals coming from the forearm are carriers of this piece of information. Useful body signals are those detected by surface Electromyography and Mechanomyography. After a brief description of the process of production of this signals, we will analyze the two methods in order to understand which characteristics could be useful to our goals.

4.2.1 Introductive Description of Body Signals: Production Process

Both EMG and MMG signals are produced by the activation of motor units. Each motor unit consists of an α-motor neuron and of the muscles fibres it innervates. When muscles contract, an increasing number of motor units are activated, by means of the signal of each motor neuron to its fibres. This process is called motor unit recruitment and accomplishes the increasing gradation of contractile strength. EMG and MMG acquire the interference signal, electrical or mechanical, of the activity of more motor units.

The relationship between the applied force and this signal is evident, being the command given to muscle contractions. Less obvious is the relationship with fatigue. The well-known production of lactates in condition of fatigue causes a change in intracellular pH and, consequently, a decrease in muscle fibre conduction velocity. The latter affects the motor unit action potential waveform, that is observable by means of EMG and MMG signals. As intuition suggests, a decrease in median frequency will be detectable in both signals.

4.2.2 Description EMG

4.2.2.1 Physical Concept

Surface Electromyography measures the interference signal of Motor Units Active Potential, that is the means of communication of motor unit components. The voltage acquired by EMG is exploited by calculating its RMS and Mean Frequency. The energy in the signal, estimated by RMS, is directly linked to the firing rate and number of motor units activated. This two parameters are related, as explained before, to the muscle contraction strength: for this reason, RMS is a good indicator of the force applied by the hand.

Mean frequency is used to detect a state of fatigue, because it is a good estimator of lactates concentration in muscles' volume.

Analyzing tests made in pre-fatigue and fatigue condition, it is clear that RMS it is not affected by this variable, whereas, as expected, a decrease in mean frequency is observed (Claudio et al. 2003) (Fig. 4.8).

4.2.2.2 Protocol: Technical Considerations

EMG signals are detected by a number of electrodes put on the muscle surface. Their range its between 50 μV and 30 mV with a frequency of 40–80 Hz.

Both MMG and EMG must be put in precise positions, but EMG requires specific attention as far as skin condition is concerned. Sweat is particularly critical and will be further investigated in the following sections.

As we observed in the previous paragraph, EMG signals can be used to detect both the strength the astronaut wants to apply and his fatigue condition.

Reliability on EMG signal for feature recognition and extraction has been proved by several studies. For example with ten electrodes twelve finger/hand postures were discriminated (van der Smagt et al. 2009).

Obviously, EMG signals have to be conditioned to be analyzed: amplification needs to be implemented, by means of a low pass filter and sampling at 1000 Hz.

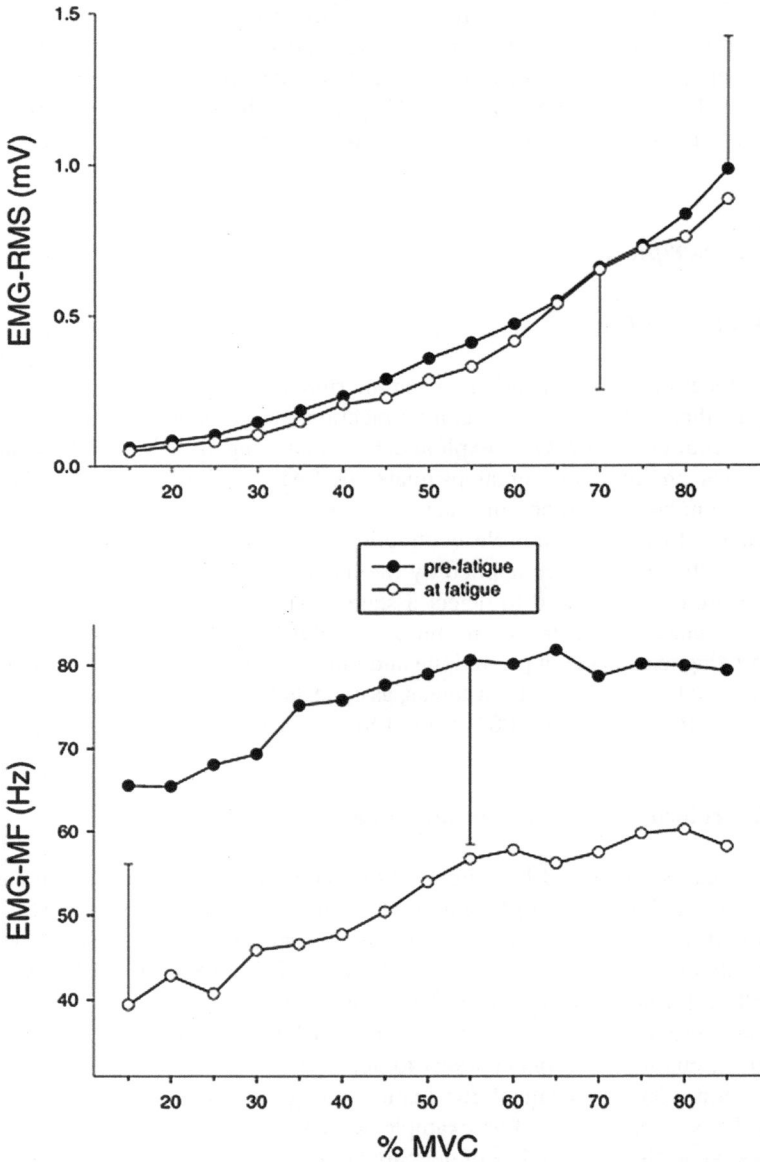

Fig. 4.8 EMG root mean squares and mean frequency dependence on maximum voluntary contraction percentage and fatigue condition

4.2.2.3 Advantages EMG

The use of sEMG has a big advantage: it is a mature technology. Research on force-electrical signals relations is available in the literature; the protocol of electrodes choices, number and positioning are optimized and well-defined; limits of environmental conditions in which the measures are reliable are already known.

Moreover, sEMG signals are relatively high frequency signals, allowing for a more rapid control actuation.

The signal RMS is not affected from fatigue condition, thus its *reproducibility* with respect to this factor is verified; it can be used as a correct unbiased estimator of the force the astronaut wants to reach. Root Mean Squares of a signal is easy to calculate and the window used for calculation can be chosen in order to minimize the noise of the measure.

4.2.2.4 Disadvantages EMG

Surface EMG laboratory experiments usually require the skin to be prepared with an alcohol swab before recordings, to improve skin conductance. In EVAs it is difficult to maintain an alcohol film on the skin, therefore the training and the real operating condition will not permit to obtain the best potential performances.

Skin conductivity highly changes with sweat. The accuracy and reliability of EMG signals are negatively affected by sweat. In fact, the formation of a sweat layer under the electrodes separates them from the surface of the skin, which in turn disrupts the quality of the resulting signal. In particular, the amplitude of EMG signals suffers a decrease of 3 % per 0.02 mm of sweat underlying the electrodes (conservative estimation) (Abdoli-Eramaki et al. 2012).

Because the signal is disrupted even with relatively small quantities of sweat, this drawback represents a serious problem in the field of application of the project; in fact, astronauts are subject to a considerable amount of sweating during EVAs. In Abdoli-Eramaki et al. (2012) it is noted that the application of medical adhesive under the electrodes practically cancels the negative effect of sweat. However, the effectiveness of such solution in the environment present inside the EVA suit needs to be confirmed. This is the most critical problem of the use of sEMG.

4.2.3 Description MMG

4.2.3.1 Physical Concept

The mechanomyogram (MMG) is the recording of mechanical oscillations associated with the muscle contraction. Oscillations are produced by the lateral dimensional changes in active muscle fibers that generate pressure waves.

MMG RMS increases with recruitment of MUs, thus with muscle strength, but at high contraction level (Maximum Voluntary contraction >65 %, when complete recruitment happens) the increasing firing rate brings to reduce the RMS in pre-fatigue conditions.

In fatigue conditions the RMS behavior is completely different, whereas the median frequency information maintains the same trend with a shift towards smaller values. Again the theory on the decrease of conduction velocity and therefore of the mean frequency with fatigue is confirmed (Fig. 4.9).

4.2.3.2 Protocol: Technical Considerations

MMG, like EMG, is detectable on the skin's surface, but it can be acquired by means of a simple accelerometer or a microphone. Acceleration range is between 0 and 1.5 m/s^2 and bandwidth is 10–40 Hz. Reliability of MMG has been proved by several studies (Xie et al. 2009), where, as an instance, two channels of MMG signals allowed to differentiate four different motions of forearm and hand.

4.2.3.3 Advantages MMG

Since the MMG is a mechanical signal, there are two main advantages. First of all no skin preparation is required to obtain the best performance. This means that sweat condition is completely uncorrelated to output signal and quality of measure is not affected. Moreover, the signal is already an average taking into account the actual effect of the Motor Units action, as a partial fusion of single events and not as the summation of distinct MUAPs as in the sEMG (Claudio et al. 2003).

It does not require electrodes to be mounted on people's skin: MMG is detected using a very small accelerometer, which is attached to the surface of the skin. Regarding the "quality" of the measure, the signal-to-noise ratio of MMG is typically higher than that of EMG.

Finally, since the accelerometer records movements directly, motion artifacts such as unwanted limb movements produce sudden large peaks in the signal, which can easily be detected and ignored.

4.2.3.4 Disadvantages MMG

The main advantage of sEMG is also the main disadvantage of MMG: in fact properties and applications have yet not been well explored, because they are really innovative technologies. Very little information is available for MMG about positioning of accelerometers, measures repeatability and optimized conditioning process. Experiments and trainings have to be performed in order to overcome this problem.

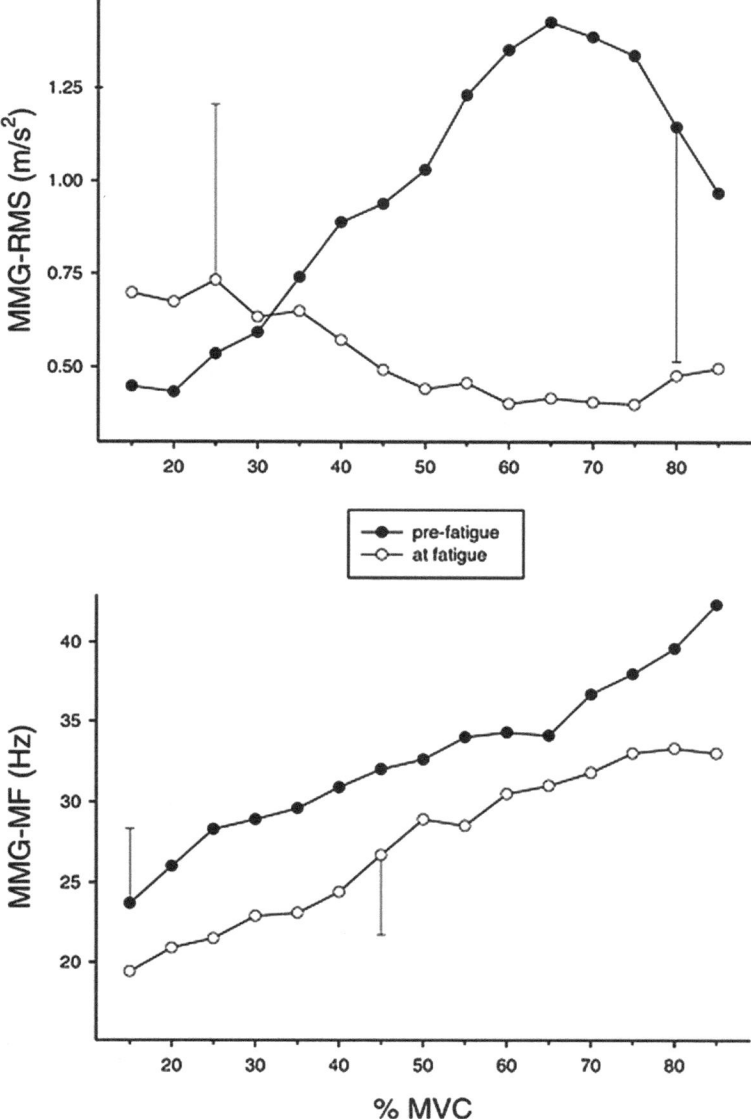

Fig. 4.9 MMG root mean squares and mean frequency dependence on maximum voluntary contraction percentage and fatigue condition

The frequency of the signal is lower than the sEMG's one (10–40 Hz), thus control actuation will be definitely slower.

As it was shown before, the RMS of MMG is highly dependent by fatigue conditions, does not satisfy the requirement of repeatability and, therefore, cannot

Table 4.5 The table sums up the main characteristics analyzed to give a general overview of the comparing process

	sEMG	MMG
Maturity of technology	✔	×
Number of detectable features	✔	×
Skin condition independence	×	✔
Reproducibility with sweat	×	✔
Reproducibility with fatigue	✔	×
Signal to noise ratio	×	✔
Speed of control	✔	×
Required calculations	✔	×

be used. As a consequence, we have to use the frequency of the signal. This has two main critical consequences; first, frequency signal estimation requires more demanding calculations for the processor, secondly it sets an inferior limit to the windowing of signal, implying a delay in the control strategy. If the minimum frequency signal is about 10 Hz, in order to recognize the smallest frequencies we have to window the signal at 0.1 s, causing a mean delay in the control of 0.05 s.

4.2.4 Sum Up Table and Critical Considerations

Obviously, some factors are more influent than others. For example, reproducibility with sweat and fatigue has two different differential gains with respect to the two factors; sEMG in conditions of sweat higher than a threshold will stop to be reliable, whereas MMG does not show reproducibility with respect to fatigue only for RMS signal. Therefore, the use of MMG mean frequency signal could overcome the problem of fatigue, with a loss in the information obtainable compared to the base case (from force and fatigue state measured by sEMG, to only force measured by MMG). The reproducibility with respect to sweat completely undoes the chance of using sEMG.

Lower importance is attributed to S/N ratio and skin condition independence (here sweat factor is excluded) because of the actually small decrease of performances of the measure system caused by these two factors. See Table 4.5 for further information.

4.2.5 Measure System Concept and Algorithm

The final measure system was implemented focusing on *redundancy, safety control* and to assure a minimum performances also in off-nominal conditions. We decided to exploit the best characteristic of the two paradigms, to obtain the best result.

We split therefore the algorithm of functioning in three different situations, depending on the matching between command signals given from the control and EMG-MMG signals and feedback sensors measures on the fingers. If the matching is not properly verified we say we are in *abnormal conditions.*

4.2.5.1 Case A

This case is the nominal one, when every sensor is properly functioning. The best performance and more reliable system to identify the user's will is employed, that is EMG RMS. The measure of the MUAPs is demanded to EMG, in order to extract the feature the astronaut wants to perform and to detect the amount of force to be sustained. Instead, fatigue status is identified combining EMG and MMG mean frequency and comparing them to the ones corresponding to pre-fatigue and at-fatigue values at the actual value of MVC (maximum voluntary contraction) percentage applied. This method derives from a research demonstrating that combining MMG and EMG signals is the most useful method to discriminate the fatigue condition from the pre-fatigue condition because the power of statistical test increases (Ioi et al. 2006).

Sensors of torque applied to the fingers, developed with the same material of actuation system, as we will describe in following sections, close the control loop giving the feedback for the control and assuring that torque would not exceed safety limits.

Sensors of position are used to control that the hand will not exceed prescribed safety limits and that performed movements are the ones corresponding to feature extracted by EMG signals processing.

If safety limits are exceeded, the algorithm goes to Case C, whereas if it the correct functioning of EMG control is not verified (which can happen when sweat quantity is far from the nominal condition), the algorithm goes to Case B.

4.2.5.2 Case B

This case is the low performance situation. Once it has been detected that EMG signals are not properly controlling the hand, passing to MMG signals would overcome problems deriving from skin condition (sweat, other changes from nominal characteristics) assuring a lower performance, but reliable, control. In fact, in the actual state of research, MMG can provide a lower amount of information than EMG, but is reliable enough to perform simple tasks. Its operation could be necessary in case of high fatigue and important but simple tasks to be performed.

In this case, the EMG control is deactivated, while MMG mean frequency values are used to control the hand. It is necessary to use the mean frequency values instead of RMS because, as we described before, RMS changes abruptly its behavior in conditions of pre-fatigue or at fatigue. On the other hand, mean frequency is subjected to a constant shift only, so it can provide reliable information, even though not exactly at the correct force, because the close control loop will bring to the willed force.

The tasks performed by position and force sensors are exactly the same as before. Obviously, in the case of incorrect functioning of MMG control, the algorithm passes to Case C.

4.2.5.3 Case C

The third case is thought to allow free motion to the astronaut, so that he will not receive help neither contrasting action from the glove. In case control, for some reasons, is not functioning properly, it could bring hard problems, on safety and tasks performing. Therefore, deactivating it would solve this problem.

The low Young modulus of the Elastomer would be the only resistance to be overcome, so that this will not be a big effort for the astronaut.

4.3 Control System Architecture

The layout of the proposed concept is represented in Fig. 4.10:

The arm on the left represents the user's forearm; EMG and MMG signals are used as input for the exoskeleton. This way, the control interface is virtually invisible and seamless as the user himself is not required any effort to provide inputs to the device. Therefore, the physiological signals will provide a "reference signal" to the controller and the latter will take care of actuating the exoskeleton to accomplish the desired movement.

Once the controller has actuated the device, it is fundamental to understand which is the current position of the fingers of the exoskeleton. This is really important as the device should ensure a high degree of safety for the user (i.e. it must avoid that the user's fingers are crushed by too much force once that the fist is completely clenched).

The actuation units, the feedback sensors and the structure of the final device itself will be combined in the same component: the special materials here proposed to build the exoskeleton, known as electroactive polymers, can be used to accomplish all the previous tasks. This will allow to satisfy many requirements at the same time: the bulkiness and the total weight of the final device will be greatly reduced, the movements of the natural hand will not be hampered and the device itself could fit inside the astronaut's glove.

The overall control strategy should rely on the previous pieces to accomplish the task of controlling the whole system. In particular, the controller unit will be in charge of:

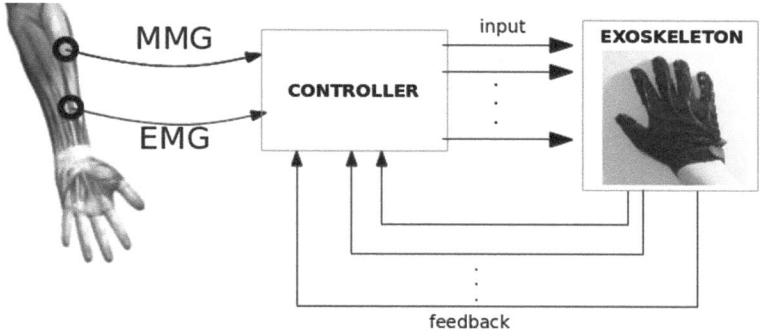

Fig. 4.10 Control system architecture

- *Reading and processing* EMG and MMG signals on the user's forearm. This step involves the acquisition of the physiological signals from the sensors and their electrical processing (amplification and noise filtering).
- *Decoding* of the acquired signals. Once the physiological signals have been processed, they must be decoded so that they can be translated into input signals for the exoskeleton.
- *Actuating* the device. A proper actuation strategy should be chosen in order to actuate the polymers lying on the device and to obtain the desired movement.
- Acquiring the *feedback* signal from the exoskeleton. The feedback signals coming from the exoskeleton must be continuously read, both to make sure that the desired position/force is achieved on each finger, and for security reasons (if the exoskeleton is not responding to the input signals, the controller must take care of deactivating it).

The decoding phase is implemented by a software component called "neural network". After an initial phase of training, this will be able to read the input EMG/MMG signals and "understand" which is the corresponding movement to be actuated on the exoskeleton (Fig. 4.11).

The training phase is done just once ("off-line") to ensure that the control algorithm will be able to recognize all the inputs provided during the "on-line" functioning (however, sometimes it could be useful to repeat this phase to "recalibrate" the control algorithm).

The training phase consists in giving the neural network a set of inputs (the EMG/MMG signals in our case) and the corresponding outputs (the fingers' positions). At this regard, gloves that contain electrogoniometers, accelerometers, flex sensors or other sensors are used to determine which is the absolute position for each finger. In our case, we plan to use the same exoskeleton, as the materials there contained can be used as sensors to detect the fingers' positions.

After the input and output signals have been recorded, the neural network will use the so called "back propagation" algorithms to define a set of features to classify the input signals. These sets are defined in multi-dimensional spaces defined on variables as the frequency of the signals, their amplitude, their spatial

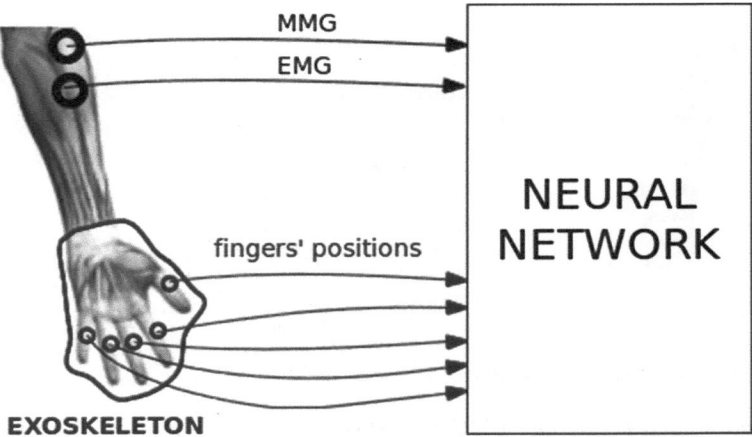

Fig. 4.11 Neural network training process

origin on the forearm and many others. The output of the training phase is the definition of a correspondence between these sets and the outputs (i.e. the movement of each finger or the hand conformation).

After the training phase, the neural network can be used "on-line", as a classifier; this means that the control algorithm will cyclically acquire the physiological signals from the sensors and use the previously defined sets to classify them. In order for this to happen, the input signals will be analyzed and their main features will be extracted; then, it will be possible to represent the input signal as a point in the multidimensional space previously defined and the classifier could then choose which is the set that best matches this point. Once the set has been found, the algorithm looks for its correspondent movement of the arm and the motors can be actuated.

4.3.1 Neural Network Training Protocol

Because the core of our movement recognition strategies relies on neural networks, a training protocol was established for their initial training. There are three main movements that must be recognized:

1. Single finger flexion (thumb, index and middle finger);
2. Pinching (Thumb-index and Thumb-Middle);
3. Full hand grasping;

For each of these movements the following steps must be performed:

1. Start from a relaxed position (Finger extended/Hand open)
2. Flexion/Contraction (*Active phase*)

3. Pause for 2 s
4. Return to the initial position

These steps should be performed ideally tens of times, at different execution speeds and by different subjects. The pause at the middle of the movement is inserted so that it is easier to distinguish the two phases of the movement (active phase and return to the initial position).

References

Abdoli-Eramaki M, Damecour C, Christenson J, Stevenson J (2012) The effect of perspiration on the sEMG amplitude and power spectrum. J Electromyogr Kinesiol 22(6):908–913

Aliev AE et al (2009) Giant-stroke, superelastic carbon nanotube aerogel muscles. Science 323:1575–1578

Bar-Cohen Y (2002) Electro-active polymers: current capabilities and challenges. San Diego, CA

Brochu P, Pei Q (2010) Advances in dielectric elastomers for actuators and artificial muscles. Macromol Rapid Commun 31:10–36

Carpi F, Kornbluh R, Sommer-Larsen P, Alici G (2011) Electroactive polymer actuators as artificial muscles: are the ready for bioinspired applications? Bioinspiration & Biomimetics 6(4):040201

Claudio O et al (2003) The surface mechanomyogram as a tool to describe the influence of fatigue on biceps brachii motor unit activation strategy. Historical basis and novel evidence. Eur J Appl Physiol 90:326–336

Ioi H et al (2006) Mechanomyogram and electromyogram analyses for investigating human masseter muscle fatigue. Orthodontic Waves 65(1):15–20

Madden JDW et al (2004) Artificial muscle technology: physical principles and naval prospects. IEEE J Oceanic Eng 29(3):706–728

Pelrine R, Kornbluh R, Joseph J (1998) Electrostriction of polymer dielectrics with compliant electrodes as a mean of actuation. Sens Actuators A 64:77–85

Pelrine R, et al (2000a) High-field deformation of elastomeric dielectrics for actuators. Mater Sci Eng C(11):89–100

Pelrine R, Kornbluh R, Pei Q, Joseph J (2000b) High-speed electrically actuated elastomers with strain greater than 100 %. Science 287:836–839

Van der Smagt P et al (2009) Robotics of human movements. J Physiol 103(3–5):119–132

Xie H, Zheng Y, Guo J (2009) Classification of the mechanomyogram signal using a wavelet packet transform and singular value decomposition for multifunction prosthesis control. Physiol Meas 30(5):441–457

Chapter 5
Concept Layout

Abstract The chapter presents the design of the device actuators and sensors in terms of concept developing, dimensioning, testing and prototyping. Inspired by soft robotics concept, the design of actuators is based on the characteristics of dielectric elastomers and their relationship between voltage applied and material elongation. Smart material properties are exploited in the multiple layers actuator proposed, capable of performing both linear elongation and bending, by means of the application of different voltage levels to different layers. Calculations of the torques required by the hand actions and, consequently, the maximum stress are presented. Then, the choice of the material leads to the identification of the correct dimensioning of the system, in terms of voltage to be applied and thick of the actuator layers. In order to control the actuation according to the human will, sensors based on physiological signals are required. EMG and MMG based sensors are chosen, specifying the sensors number and positioning required for an effective human will identification. Pressure sensors, realized with dielectric elastomers, give the feedback of the system. Calculations relating voltage measured and pressure applied on the sensors are presented. Finally, the prototyping and testing phase is described, in which two mockup hands are realized for a proof of concept while EMG and MMG sensors are tested. Experimental setup and qualitative results are extensively described.

In the previous sections, all the technological solutions needed for the hand glove have been described and the most suitable ones have been identified. For the integration of the different aspects of the solution, we focused on the index finger.

Chapter written with the contribution of Roberto Rossi (Politecnico of Milano).

P. Freni et al., *Innovative Hand Exoskeleton Design for Extravehicular Activities in Space*, PoliMI SpringerBriefs, DOI: 10.1007/978-3-319-03958-9_5, © The Author(s) 2014

5.1 Actuator Design

The actuation device is based on the selected electroactive polymers, the dielectric elastomers. DEs proved to be the best trade-off in terms of deformability and maximum stress they are able to apply. According to their working principle, these smart polymeric materials are linear actuators, i.e. they apply a stress along a plane, mainly in one direction (if prestrained). On the other hand, the finger joints need a bending actuation and, therefore, an appropriate design of the device becomes necessary.

The adopted solution is based on the unimorph cantilever working principle: the coupling of a DEs layer with a passive layer results in an overall bending of the device (Fig. 5.1).

Due to the peculiarity of finger joints, aside from bending, the device needs also to linearly elongate. For this reason it is not possible to use a passive stiff layer in the unimorph design.

Consequently the multilayer structure designed is made of N active layers. In the example presented in Fig. 5.2 $N = 5$. Each layer performs a different linear elongation, so that $\varepsilon_5 > \varepsilon_4 > \varepsilon_3 > \varepsilon_2 > \varepsilon_1$. The resulting actuation is both a linear expansion along the finger and a bending on the joint.

To achieve a gradient of linear strains across the device's thickness, a different voltage is applied on each layer. According to the considerations drawn in Sect. 4.1.4, the planar strain along the polymeric dielectric increases with the bias applied and, therefore

$$\varepsilon_5 > \varepsilon_4 > \varepsilon_3 > \varepsilon_2 > \varepsilon_1 \Leftrightarrow V_5 > V_4 > V_3 > V_2 > V_1$$

As a result, applying the proper voltage to each layer, the device actuates according to the requirements, remaining compliant with the finger. The device described above has been named Multilayer Elongation and Bending Actuator (MEBA).

Figure 5.3 shows the arrangement on the index finger of the three MEBAs, one for each joint (MCP in green, PIP in red and DIP in yellow). Every MEBA needs to be dimensioned according to the specific geometry of the joint and to the loading conditions.

5.2 Kinematics and Dynamics Specifications

The needed actuation obviously depends on the applied external forces and on the way these forces are translated to torques on finger joints. After having identified the torque on each joint, the interaction between finger and actuation and the geometrical and physical characteristics of the Elastomer, it is possible to define the maximum stress acting on Elastomer. This is also the stress the Elastomer must provide by means of voltage-stress relationship; thus, we will verify the coherence between available actuation stress and required stress.

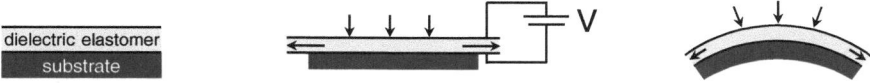

Fig. 5.1   Unimorph DE actuators working principle

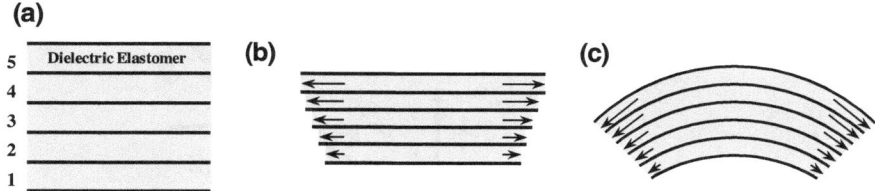

Fig. 5.2   Multilayer Elongation and Bending Actuator (MEBA). **a** No voltage applied. **b** The application of different voltages causes different linear elongations. **c** The resulting deformation is both a linear elongation and a bending

Fig. 5.3   MEBAs arrangement on the index finger

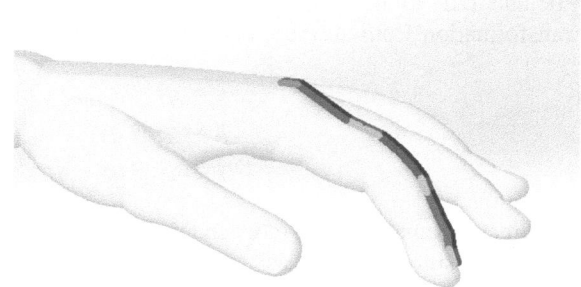

Fig. 5.4   Joints of finger

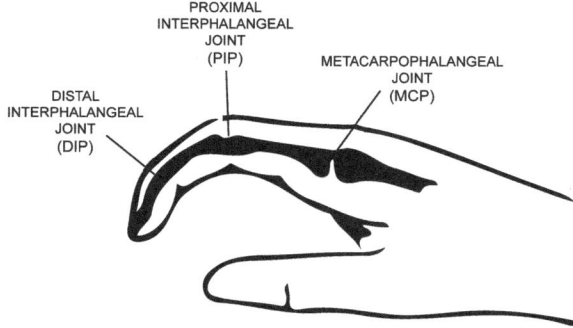

In order to draw the dynamical specifications for the actuation of the finger we approximated the three joints (Fig. 5.4) to three hinges and we used this simplified but effective model (Fig. 5.5).

Fig. 5.5 Dynamic model of
the finger

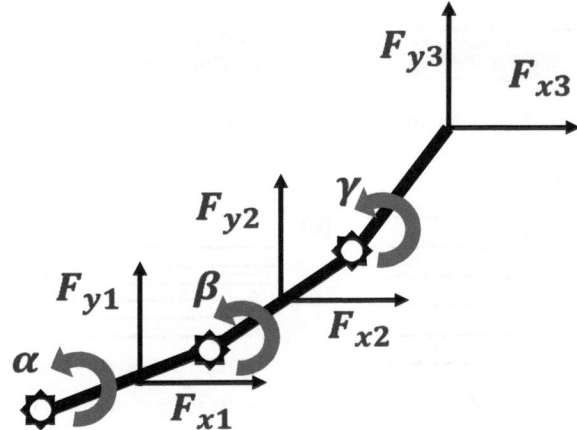

The three coordinates α, β, γ are respectively the absolute angles of the MCP, PIP and DIP joints. The kinematic model is easily obtainable with the useful transformation from absolute coordinates to relative ones.

$$\alpha = q_1$$
$$\beta = q_2 + q_1$$
$$\gamma = q_3 + q_2 + q_1$$

Direct kinematics from independent and relative coordinates to distal phalange absolute position is found as:

$$\begin{cases} x_3 = x_0 + l_1\cos(q_1) + l_2\cos(q_1 + q_2) + l_3\cos(q_1 + q_2 + q_3) \\ y_3 = y_0 + l_1\sec(q_1) + l_2\sec(q_1 + q_2) + l_3\sec(q_1 + q_2 + q_3) \end{cases}$$

Calculating the derivatives of these two equations, we can find the Jacobian, to obtain this form.

$$\dot{x} = \mathrm{J}(q) \cdot \dot{q}$$

The Jacobian is calculated as:

$$\mathrm{J}_3(q) = \begin{bmatrix} -l_1\mathrm{s}(q_1) - l_2\mathrm{s}(q_1 + q_2) - l_3\mathrm{s}(q_1 + q_2 + q_3) & -l_2\mathrm{s}(q_1 + q_2) - l_3\mathrm{s}(q_1 + q_2 + q_3) & -l_3\mathrm{s}(q_1 + q_2 + q_3) \\ l_1\mathrm{c}(q_1) + l_2\mathrm{c}(q_1 + q_2) + l_3\mathrm{c}(q_1 + q_2 + q_3) & l_2\mathrm{c}(q_1 + q_2) + l_3\mathrm{c}(q_1 + q_2 + q_3) & l_3\mathrm{c}(q_1 + q_2 + q_3) \end{bmatrix}$$

In the same way we can calculate the Jacobian between relative joints and the position of Medial Phalange and Proximal Phalange.

Motions	Force (N)
Table 5.1 Forces of the finger joints (Favetto et al. 2010)	
Finger MCP joint	3.92
Finger PIP and DIP joint	8.82
Lateral motion of thumb	7.84
Thumb MCP joint	6.86
Thumb PIP and DIP joints	5.39

$$J_2(\boldsymbol{q}) = \begin{bmatrix} -l_1\mathrm{s}(q_1) - l_2\mathrm{s}(q_1 + q_2) & -\frac{l_2}{2}\mathrm{s}(q_1 + q_2) & 0 \\ l_1\mathrm{c}(q_1) + l_2\mathrm{c}(q_1 + q_2) & \frac{l_2}{2}\mathrm{c}(q_1 + q_2) & 0 \end{bmatrix}$$

$$J_1(\boldsymbol{q}) = \begin{bmatrix} -\frac{l_1}{2}\mathrm{s}(q_1) & 0 & 0 \\ \frac{l_1}{2}\mathrm{c}(q_1) & 0 & 0 \end{bmatrix}$$

Once the three Jacobians obtained, it is possible to calculate the effect of contact forces on the finger in the shape of joint torques that will be actuated.

Several studies (Favetto et al. 2010) have been performed on hand-tools interactions to identify the forces applied by the hand in actions similar to the ones that astronauts do in EVAs. Obviously, we start from an experimentally identified set of forces and we multiply them for a coefficient to assure that the actuation can provide the maximum required force (Table 5.1).

The needed torques are calculated,

$$\begin{Bmatrix} \tau_1 \\ \tau_2 \\ \tau_3 \end{Bmatrix} = J_1(\boldsymbol{q})^T \cdot \begin{Bmatrix} F_{x1} \\ F_{y1} \end{Bmatrix}, \quad \begin{Bmatrix} \tau_1 \\ \tau_2 \\ \tau_3 \end{Bmatrix} = J_2(\boldsymbol{q})^T \cdot \begin{Bmatrix} F_{x2} \\ F_{y2} \end{Bmatrix}, \quad \begin{Bmatrix} \tau_1 \\ \tau_2 \\ \tau_3 \end{Bmatrix} = J_3(\boldsymbol{q})^T \cdot \begin{Bmatrix} F_{x3} \\ F_{y3} \end{Bmatrix}$$

And their sum give us the total torques.

$$\begin{Bmatrix} \tau_1 \\ \tau_2 \\ \tau_3 \end{Bmatrix} = J_1(\boldsymbol{q})^T \cdot \begin{Bmatrix} F_{x1} \\ F_{y1} \end{Bmatrix} + J_2(\boldsymbol{q})^T \cdot \begin{Bmatrix} F_{x2} \\ F_{y2} \end{Bmatrix} + J_3(\boldsymbol{q})^T \cdot \begin{Bmatrix} F_{x3} \\ F_{y3} \end{Bmatrix}$$

The calculations for all the positions are necessary to identify the maximum absolute torques, the maximum torque required at maximum rotation, and to understand if there are positions in which specifications are less strict.

With rough calculation is easy to find, in the most critical configuration, the following values of torques. We assume as forces values the ones reported in the table in order to have the chance to totally sustain the external loads by means of the robotic glove.

$$\begin{Bmatrix} \tau_1 \\ \tau_2 \\ \tau_3 \end{Bmatrix} = J_1(0)^T \cdot \begin{Bmatrix} 0 \\ 3.92\,N \end{Bmatrix} + J_2(0)^T \cdot \begin{Bmatrix} 0 \\ 8.82\,N \end{Bmatrix} + J_3(0)^T \cdot \begin{Bmatrix} 0 \\ 8.82\,N \end{Bmatrix}$$

$$\begin{Bmatrix} \tau_1 \\ \tau_2 \\ \tau_3 \end{Bmatrix} = \begin{Bmatrix} 7.9 \cdot 10^{-1}\,Nm \\ 2.4 \cdot 10^{-1}\,Nm \\ 4.4 \cdot 10^{-2}\,Nm \end{Bmatrix}$$

These approximate values are useful to evaluate the feasibility of the solutions we analyze. We remind that the torque τ_1 calculated in this way is referring to the joint MCP, the torque τ_2 to the PIP and τ_3 to DIP.

5.3 Structural Model of the Actuator

Actuation is provided by the three MEBAs put on the upper side of the finger, each one corresponding to a joint. The functioning of the three joints is analogue, thus we can analyze only one of them. In particular the PIP joint was considered; with appropriate changes, the found solution can be adapted to the other joints of the finger.

The actuator, in contact with the end, exchanges a train of loads to maintain the finger in equilibrium state or to provide a rotation to it. A first modeling can be the one depicted in Fig. 5.6. The momentum reported is caused by the external forces, whose relationship and value are explained above (Sect. 5.2) (Fig. 5.7).

The two elements are grounded because the finger is joined to the hand and the actuator is joined to the previous one, blocked by the control. In order to consider the worst situation, we substitute the pressure load train with a concentrated force at the end of the contact area between the actuator and the finger (Fig. 5.8).

The maximum momentum the actuator has to sustain is given by the simple expression:

$$M_{max} = \tau \cdot \frac{l_1}{l_2} = 2\tau$$

From now on, we will focus on the second joint (PIP), because the CMC is in a position allowing more space for the actuator, so that it would not be a problem to increase the dimension of the MEBA to sustain a larger momentum. The DIP joint has such a small torque that dimensioning is not critical.

The maximum mechanical stress required to apply this momentum, referring to the structure of MEBA, and with the reasonable hypothesis that the number of elements it is large enough, could be calculated this way:

$$M_{max} = b \cdot \sum_{1=i}^{N} \sigma_i \cdot y_i \cdot h_i = b \cdot \int_{-\frac{t}{2}}^{\frac{t}{2}} \sigma \cdot y \cdot dy$$

In which b is the width, y is the vertical coordinate (with the origin in the middle of the section), and h_i is the thickness of each layer. The MEBA is actuated to have a

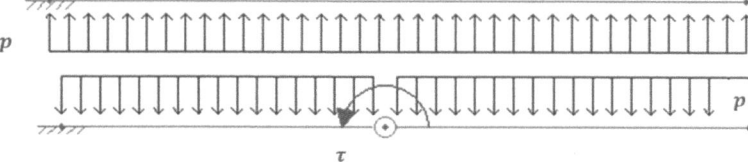

Fig. 5.6 Interaction model between actuator (*top*) and finger (*bottom*)

Fig. 5.7 Visual
representation of the
interaction between actuator
and finger

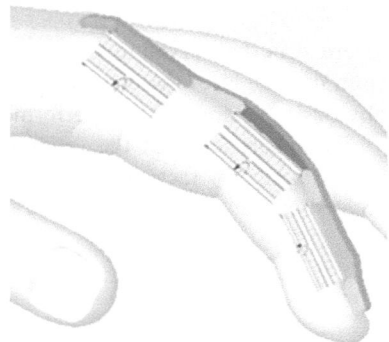

linear distribution of σ, written as the sum of a mean value σ_t, and a linear zero
mean component, $\sigma_f = \sigma_{max,f} \cdot \dfrac{y}{t/2}$, where t is the thickness of the MEBA.

$$M_{max} = b \cdot \int_{-\frac{t}{2}}^{\frac{t}{2}} \left(\sigma_t + \sigma_{max,f} \cdot \frac{y}{t/2} \right) \cdot y \cdot dy = \sigma_{max,f} \cdot b \cdot \frac{t^2}{6}$$

$$\sigma_{max,f} = \frac{M_{max}}{b \cdot \frac{t^2}{6}}$$

So, the maximum stress is given by the sum of the spring back effect and of the
momentum given to equilibrate the structure.

The elastic return effect depends on the rotation of the considered joint. In fact,
being the neutral axis not in the center of the MEBA, it will be subjected to a linear
deformation with mean value different from zero.

The maximum bending strain is computed from:

$$\varepsilon_{f,max} = \frac{y_{max}}{\rho} = \frac{t}{2\rho} = \frac{t}{2\left(l_\rho + \frac{t}{2}\right)}$$

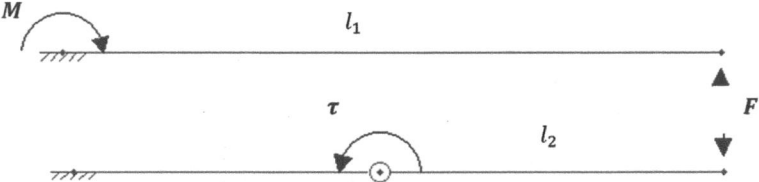

Fig. 5.8 Simplified interaction model between actuator (*top*) and finger (*bottom*)

where ρ is the curvature radius, which split in two components, one depending from the height of the material $\frac{t}{2}$ and the remaining being l_ρ. The curvature radius obviously depends on the performed rotation.

The mean strain of the section is easily computed considering the curvature radius again. In fact, the overall mean deformation is given by the arc of the created circumference and the strain is computed in the following way:

$$r \cdot \alpha = \Delta L \Longrightarrow \varepsilon_t = \frac{r \cdot \alpha}{L_0} = \frac{\left(\frac{h_f}{2} + \frac{t}{2}\right) \cdot \alpha}{L_0}$$

where L_0 is the length of the MEBA, h_f is the finger thickness, and α is the rotation of the joint. Finally, the maximum total stress to be sustained is given by:

$$\sigma_{tot} = \sigma_{max,f} + Y\left(\varepsilon_{f,max} + \varepsilon_t\right) = \frac{M_{max} \cdot 6}{b} \cdot \frac{1}{t^2} + Y \cdot \left(\frac{t}{2\left(l_\rho + \frac{t}{2}\right)} + \frac{\left(\frac{h_f}{2} + \frac{t}{2}\right) \cdot \alpha}{L_0}\right)$$

$$(5.1)$$

where Y is the Young modulus.

For substituting the numerical values we will focus on the second joint (PIP), because the MCP is in a position allowing more space for the actuator, so that it would not be a problem to increase dimension of the Elastomer to sustain an larger momentum, whereas the DIP joint has such a small torque that dimensioning is not critical.

Numerical values were extracted from different sources and methods, integrating:

- data of studies of researches, as in the case of the forces or average values of finger dimension;
- data given directly by astronauts in interviews;
- data experimentally measured and estimated.

Fig. 5.9 The figure reports
the deformed configuration
used to extract the geometric
parameters needed for the
dimensioning

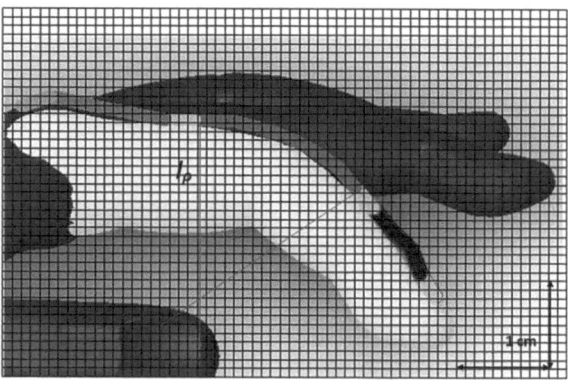

The maximum momentum to be sustained from the second joint was found to be:

$$M_{max} = 4.8 \, 10^{-1} \, \text{Nm}$$

From the interviews to astronauts it was known that the most closed condition
required is equivalent to holding a 5 cm diameter ball in one hand. It was estimated that this corresponds to a maximum rotation of 45° and to a curvature radius
of 2 cm (l_ρ) (Fig. 5.9).

The other data needed to compute the relationship between the stress σ_{tot} and
the height of the material t are:

$$Y = 3 \, \text{MPa}, \quad h_f = 0.012 \, \text{m}$$
$$l_\rho = 0.020 \, \text{m}, \quad \alpha = \frac{\pi}{4} \, \text{rad}, \quad L_0 = 0.020 \, \text{m}$$

Equation (5.1) describes the maximum stress in the top layer (the most stressed
and strained) of the MEBA. σ_{tot} can be plotted as a function of total thickness (t) of
the MEBA (Fig. 5.10): the graph shows that, as the total thickness increases, the
stress on the top layer decreases. Comparing this feature with the actuation stress
applied by the dielectric elastomer, it is possible to determine the threshold
thickness for the MEBA.

Two considerations have to be drawn to motivate the hypothesis done.

First, the high flexibility of the material could induce to think that the
momentum it has to sustain will cause too big deformation to the material itself,
changing the contact point with the finger. This phenomenon is not due to happen
because we plan to put a covering around the material, similarly to an internal
glove (Fig. 5.11), in order to prevent a radial deformation of the elastomer over a
fixed value. In this way, we will be sure that the MEBA will transmit the forces as
modeled.

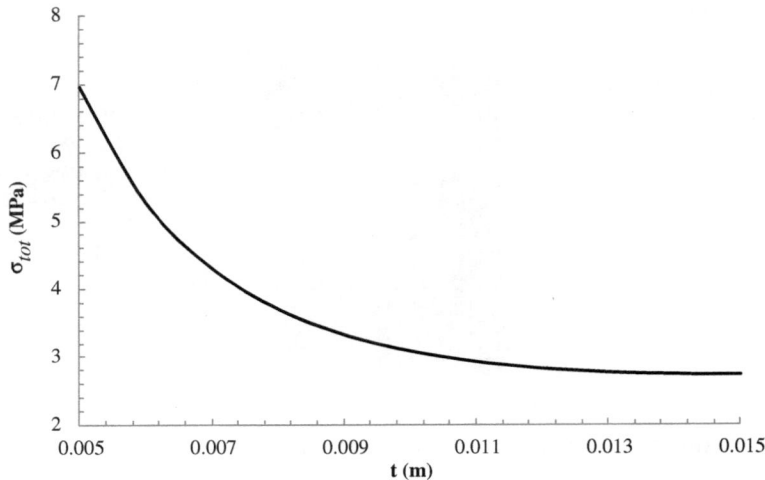

Fig. 5.10 Dependence between maximum stress on the actuator and its height

Fig. 5.11 Representation of
the external structure limiting
to a maximum radial
deformation the actuator

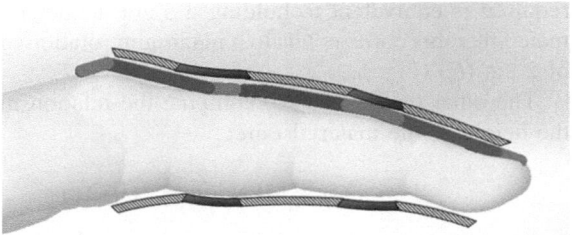

Secondly, the contact between the Elastomer and the just explained covering
will reduce the load on the most critical point of the material, increasing the safety
factor.

Lastly, the most critical conditions for the external forces and for the bending
contributions were summed. Nevertheless, they take part to two completely
opposite situation, when the finger is completely stretched for the forces, and when
the finger is completely closed for the bending. This is another reason why the
forces we can apply are underestimated in these calculations.

5.4 MEBA Dimensioning

Equation (5.1) gives an indication of the major stress that the dielectric elastomer
actuator is required to apply. Therefore, the values computed and reported in
Fig. 5.10 represent the threshold pressure that the material has to provide, as a

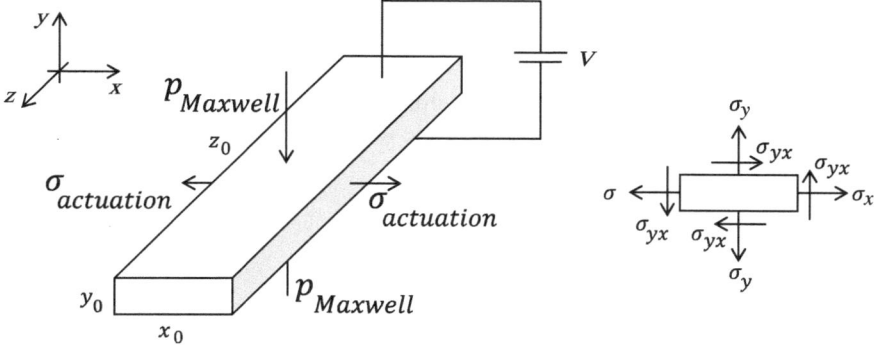

Fig. 5.12 Dielectric elastomer element

function of the device's thickness. The next step is to compute the maximum pressure the DE is able to produce and to compare it to the requirements, in order to choose the minimum thickness needed for the device to work properly.

The worst loading conditions occur in the top layer of the MEBA, so, for the following evaluations, a small element of that layer is considered. The presence of loading constraints requires a proper evaluation and, even though the mathematical model presented in Sect. 4.1.4 is no longer valid, the initial assumptions on the material are the same: we consider the DE to be incompressible, homogeneous, isotropic, with a linear elastic mechanical behavior.

The element taken into account (Fig. 5.12) is prestrained in the z-direction. This way, no further deformation along z can occur and $z_0 = const$. The Maxwell pressure, due to the application of the voltage V, is applied along y and the actuation stress is directed along x.

The loading state and the geometry of the element lead to a *plane strain* in each x–y section and it can be assumed that all the cross sections, along z, are in the same condition (Kofod 2008; Timoshenko and Goodier 1970). Taking one of those sections, the state of stress is described by the Hooke's law in the form of:

$$\begin{bmatrix} \varepsilon_x \\ \varepsilon_y \\ \varepsilon_z \end{bmatrix} = \frac{1}{Y} \cdot \begin{bmatrix} 1 & -v & -v \\ -v & 1 & -v \\ -v & -v & 1 \end{bmatrix} \cdot \begin{bmatrix} \sigma_x \\ \sigma_y \\ \sigma_z \end{bmatrix} \tag{5.2}$$

where Y is the Young modulus and v is the Poisson's ratio.

The plane strain condition imposes $\varepsilon_z = 0$ and, therefore, it is possible to write:

$$\sigma_z = v(\sigma_x + \sigma_y)$$
$$\varepsilon_x = \frac{1}{Y}\left[(1 - v^2)\sigma_x - v(1 + v)\sigma_y\right] \tag{5.3}$$
$$\varepsilon_y = \frac{1}{Y}\left[(1 - v^2)\sigma_y - v(1 + v)\sigma_x\right]$$

Moreover, no shear stress is present along x and y, therefore $\sigma_{xy} = 0$, as well as $\varepsilon_{xy} = 0$.

To evaluate the maximum actuation stress the material can apply along x, a dummy constraint is placed so that the deformation in the x-direction is imposed to be zero, $\varepsilon_x = 0$. Consequently, it is possible to write a direct relation between the stresses in the y and x directions:

$$\sigma_x = \frac{v(1+v)}{(1-v^2)} \cdot \sigma_y$$

Substituting σ_x and σ_y with $\sigma_{actuation}$ and $p_{Maxwell}$:

$$\sigma_{actuation} = \frac{v(1+v)}{(1-v^2)} \cdot p_{Maxwell} \qquad (5.4)$$

For elastomeric material the Poisson's ratio is $v \cong 0, 50$. It results, recalling Eq. (4.4), that:

$$\sigma_{actuation} \cong p_{Maxwell} = \varepsilon_r \cdot \varepsilon_0 \cdot \left(\frac{V}{h}\right)^2 \qquad (5.5)$$

where ε_r is the material relative dielectric constant, ε_0 is the vacuum dielectric constant, V is the applied voltage and h is the thickness of the DE layer.

Eventually, it is possible to express the actuation stress the material is able to exert as a function of both layer thickness and voltage applied. Superimposing the resulting curves to the plot of Fig. 5.10, it is possible to identify the working point of the actuator, dimensioning both thickness and voltage to apply.

To perform those evaluations, the Acrylic Dielectric Elastomer (3 M VHB 4910) has been chosen because it is the best performing alternative. The properties of the material are presented in Table 5.2.

It is important to point out that the maximum actuation pressure (7.2 MPa) has been computed by means of the Maxwell relation (Eq. (4.3)), applying the breakdown electric field. Therefore, this value represents the superior limit to the actuation stress that the material is able to provide, regardless the device's geometry.

The MEBA designed for the PIP joint is made of $N = 8$ layers. Thus the thickness of each layer is $h = t/8$, where t is the total thickness of the MEBA, and the actuation strain (Eq. (5.5)) becomes:

$$\sigma_{actuation} = \varepsilon_r \cdot \varepsilon_0 \cdot \left(\frac{V}{t/8}\right)^2 \qquad (5.6)$$

Table 5.2 Properties of acrylic dielectric elastomer (3M VHB 4910)

Prestrain (x % y %)	Energy density MJ/m³	Maximum actuation pressure Mpa	Young's modulus MPa	Breakdown electric field MV/m	Dielectric constant	Efficiency %	Reference
(300, 300)	3.4	7.2	3.0	412	4.8	80	(Pelrine et al. 2000)

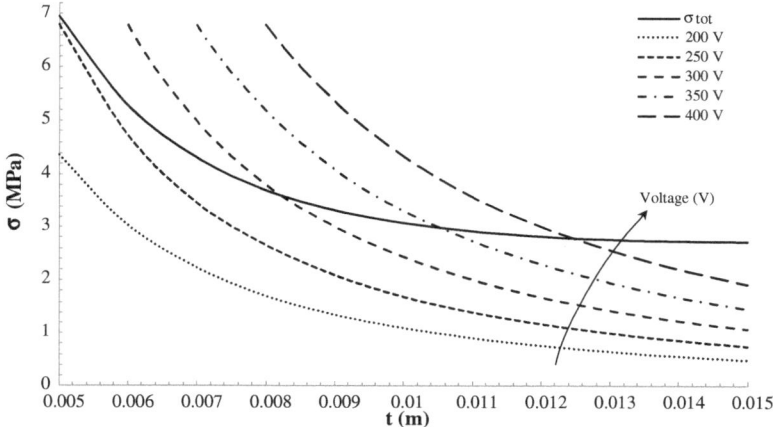

Fig. 5.13 Superposition of the actuation stresses provided by DEs at different voltages and the total stress required

In Fig. 5.13 the actuation stress, provided by the Acrylic Dielectric Elastomer on the top layer of the MEBA with different voltages, is superimposed to the total stress required on the PIP joint, as computed thanks to Eq. (5.1).

The intersections between the full line and the dotted lines identify the maximum MEBA thickness for each voltage. In order to provide enough pressure, the top layer of the MEBA needs at least the application of 300 V. With that bias, the maximum thickness of the actuator can be $t = 0.008$ m and, for the single layer, $h = 0.001$ m. Since the other layers of the MEBA are less stressed, they need a lower voltage to guarantee the proper working of the device.

According to the previous calculations and considerations, the Acrylic Dielectric Elastomer MEBA proved to satisfy the actuation requirements for the PIP joint of the index fingers. With some changes in the geometry it is possible to realize the actuators for the other joints and, eventually, a complete, light and compliant actuation for the fingers movement can be realized.

5.5 Sensors

5.5.1 EMG/MMG Sensors

A combined approach based on the use of both EMG and MMG signals was chosen for the interpretation of the user's will. In both cases a critical aspect concerns the positioning of sensors, which has to consider, first of all, the limb and hand anatomy. The team analyzed the relations among the movements of fingers and the responsible muscles, investigating the available solutions and studies found in literature.

Considering the muscular anatomy and the constraints imposed by the working conditions, the group decided to place both types of sensors on the forearm, where flexion and extension movements can be detected. Although a drawback of this choice is related to the non-direct sensing of adduction and abduction movements, which are controlled by muscles laying inside the hand, there are studies reporting the possibility of indirectly measuring their activity through the combined analysis of the forearm EMG signals (Maier and Smagt 2008).

In first approximation we can consider that flexor muscles are placed in the anterior part of the lower arm, in an anatomical compartment known as *anterior compartment* (or *flexor compartment of the forearm*) while extensor muscles are localized on the opposite side in the *posterior compartment* (or *extensor compartment of the forearm*) (Drake et al. 2009). This allows a quite feasible and reliable discrimination of flexion and extension movements, while several complications arise when an isolation of the single finger flexion and of the single finger extension is needed.

A positioning scheme for the detection of the movements of each fingers independently by means of surface EMG was proposed by Maier and Smagt (2008) and is based on the use of 10 electrodes, 3 for the extension movements and 7 for flexion movements (Maier and Smagt 2008). The scheme is illustrated in Fig. 5.14, while the main fingers' movements of interests and the responsible muscles are reported in Table 5.3.

The same approach was chosen for what concerns the MMG detection, as microphones have to be placed in correspondence to the activated muscles in order to detect the produced pressure waves. In first approximation, an easy detection of flexion and extension movements could be already obtained by placing two microphones on the anterior and posterior forearm respectively; a positioning scheme optimization is needed in order to detect and correlate the local pressure waves and the single fingers' movements.

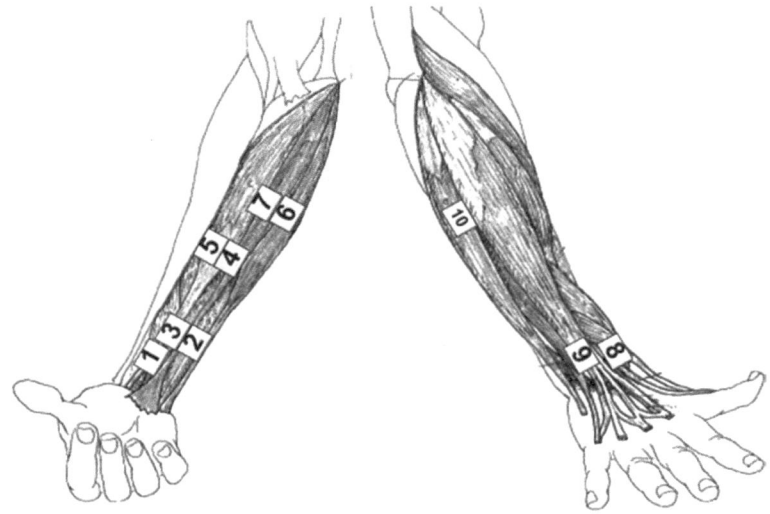

Fig. 5.14 The schematic reports a possible EMG electrodes positioning on the forearm for the detection of single fingers movements (Maier and Smagt 2008)

Table 5.3 The table lists, for each electrode, the main fingers movements to be individuated and the activated muscles considered in the detection

Electrode number	Muscle	Movement
1	Flexor pollicis longus	Flexion of the thumb
2	Flexor digitorum superficialis	Flexion of fingers, highest response for index finger
3	Flexor carpi ulnaris	Flexion of the pinkie
4–7	Fexor carpi radialis Flexor palmaris longus	Flexion of the middle and ring finger
8	Extensor pollicis longus	Extension of the thumb
9	Extensor indicis	Extension of the index
10	Extensor carpi ulnaris	Extension of middle, ring and pinkie

5.5.1.1 Feedback Sensors

To close the control loop it is necessary to have a feedback on the pressure actually applied by the fingers. For this reason, pressure sensors are needed on the fingertips of the glove.

The particular material selected to build up the actuation device shows to be suitable also for this application. In fact, Dielectric Elastomers are transducers that can be used as both actuators and pressure sensors. Recalling the working principle of DEs (Sect. 4.1.4), they can be modeled as plane capacitor with a variable capacitance. On a first approximation, the capacitance depends on the thickness of the elastomeric material. The latter can be modified applying a pressure and, therefore, monitoring the capacitance of the DE it is possible to get information about the pressure applied on it.

To translate those consideration in mathematical terms, the DE can be considered as a capacitor of capacitance C_0, charged applying a voltage V_0. This way a charge $Q = C_0 \cdot V_0$ is accumulated on the capacitor plates. Once charged, the capacitor is isolated, so that $Q = const$ is conserved.

In such a condition, a change in the capacitance of the device will cause a change in the voltage measured between the plates.

$$Q = C_0 \cdot V_0 = C \cdot V = const \Rightarrow V = \frac{C_0 \cdot V_0}{C} \tag{5.7}$$

The capacitance can be expressed according to the geometry and the material of the capacitor:

$$C = \varepsilon_r \cdot \varepsilon_0 \cdot \frac{A}{d} \tag{5.8}$$

where A is the plates surface and d is the dielectric thickness. Combining (5.7) and (5.8):

$$V = \frac{C_0 \cdot V_0}{\varepsilon_r \cdot \varepsilon_0 \cdot A} \cdot d \tag{5.9}$$

The (5.9) describe also the relation between ΔV and Δd. Moreover, recalling the Hooke's law and the definition of strain, with some trivial algebra, it is possible to write:

$$\frac{V}{V_0} = \frac{p}{Y} \tag{5.10}$$

where p is the pressure applied on the material and Y is its Young's modulus. With Eq. (5.10) it is possible to associate the intensity of the signal $\frac{V}{V_0}$ to the pressure applied on the dielectric elastomer.

5.6 Prototyping and Experiments

A great focus has been posed on the development of prototypes and mockups to test many of the ideas and concepts that the team had during the design phase.

A first important decision, for example, was related to the design of the mechanical structure: during the first iterations of the brainstorming phases, one of the most important challenges the team faced was that of thinking to a lightweight

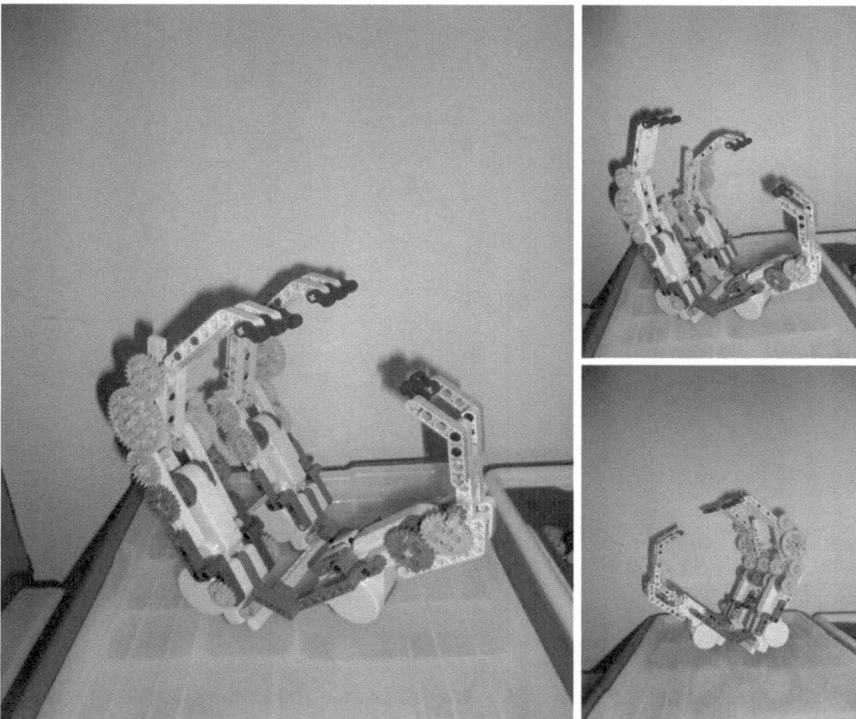

Fig. 5.15   Preliminary mockup of the mechanical design

mechanical structure capable of providing the performance required with reduced bulk, so that the device could be embedded inside astronauts' gloves.

At this regard, many alternatives were evaluated about whether to realize the structure using artificial tendons, leverages, bar mechanisms, wheels and similar mechanical designs.

One of the first mockups (Fig. 5.15) represented a hand with three fingers. The thumb was greatly simplified and designed as a body with a single link, while the other two fingers were bodies modeled with two links, imitating a system made of phalanxes.

The idea behind the design of a single finger was that of having only an actuator and a transmission system capable of moving both phalanxes at the same time but with different speeds. This was possible by tuning the diameters of the wheels and their position on the phalanxes.

Despite the kinematic chains were quite efficient, designs like this one (along with the study of the state of the art of the current exoskeletons) allowed the team to understand how this kind of mechanical structures, made of leverages, hinges and wheels (even smaller than the ones showed in the previous pictures) would

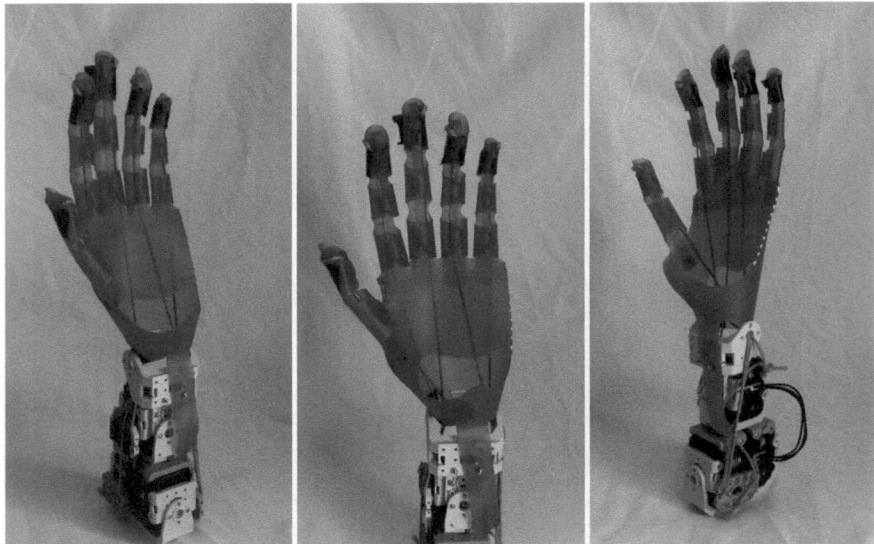

Fig. 5.16 Hand mockup for EMG experiments

have been too bulky to be embedded inside an astronaut's glove. Thanks to these first attempts, it became clear that some other solutions should be considered.

After the team decided to follow the route of "smart materials" (as described throughout the document), another important prototype was built to test some control algorithms related to the EMG signals (Sect. 5.6.1).

The mechanical structure of this prototype (Fig. 5.16) was built to allow tendons (which are clearly visible as brown ropes in the pictures above) to move each finger independently. These tendons allowed the actuation of the thumb, the index and the medium fingers by mean of three different servomotors (at the base of the hand itself).

The simple design allowed to perform some experiments related to the control of the single fingers and to the actuation of the whole hand system, giving the opportunity to develop control algorithms and test them on a device that closely resembled a real human hand.

5.6.1 Experiments with EMG

A set of experiment have been conducted in order to discover capabilities and limitations of the use of EMG signals as control signals. The goal of these experiments has been to develop a basic EMG-based control algorithm that could serve as a proof of concept for our ideas.

Fig. 5.17 Needle-based
electrodes (reproduced with
permission of
OTBioelettronica)

Fig. 5.18 Ring-based matrix
electrodes (reproduced with
permission of
OTBioelettronica)

The first step for the implementation of our initial prototype was to setup the experimental environment. Few tools were required: a set of EMG electrodes, the EMG acquisition equipment and a regular PC. Because we wanted to test our work in a real-time context, the hand mockup presented in Sect. 5.6 was used too. Most of the equipment was provided by the IIT laboratories in Torino.

Inside the laboratory, we had the possibility to test two types of EMG electrodes. The first type consisted in a set of 5 mm short needles that could be positioned on the surface of a particular muscle, in order to sense its activity (Fig. 5.17). The advantage of this solution lies in its simplicity and in the possibility of easily rearrange the sensor on a different muscle during testing.

The second type of electrodes was more but also less straightforward to be used. The electrodes were made of small metallic rings (about 5 mm in diameter) arranged in a two-dimensional matrix and insulated from each other (Fig. 5.18). The rings must be filled with a special conductive paste, and positioned on the surface of the muscle to sense. This kind of electrodes can be used only once, and they cannot be rearranged after the initial positioning.

Our experiments showed that the electrodes must be placed within a 5–6 mm range with respect to their appropriate position, and the skin should be ideally shaved and moist. For the limited scope of our initial experiments, a single set of electrodes of the first type was used (mainly for its ease of use).

The signals produced by the EMG electrodes were captured by the EMG acquisition equipment (*EMG-USB2* by OTBioelettronica (OTBioelettronica 2011)), which converts them into a dedicated digital format to be processed by the PC. However, the electrodes alone are not enough to properly sense EMG signals. Because EMG signals range in the order of mV, the EMG acquisition equipment requires also a reference voltage in order to filter interferences and spurious frequencies. Even the smallest interferences due to fluctuations in the electrical network can seriously affect the quality of the sensed signals.

Once that the acquisition environment was correctly set-up, the signals produced by the EMG acquisition equipment were processed in MATLAB thanks to a "software-bridge" (distributed by OTBioelettronica) called *OT Connector*. Because the robotic hand mockup could also be controlled through MATLAB, we focused our efforts in developing a basic hand movement recognition algorithm. After a few design iterations, we understood that the grasping movement could be easily recognized by placing an electrode on the *"flexor digitorum profundus"*. The resulting EMG signals passes through a weighted average and then is scaled in the range 0–100, where 0 correspond to fully extended fingers and 100 to fully clenched fist (a few measurements are necessary in order to associate the value of the weighted average with the different finger positions).

Once that the grasping movement was correctly recognized, we developed a software layer for converting hand-level commands (e.g. "close fist", "open hand", "flex index finger") to the commands of the individual motors that allow our mockup robotic hand to move.

At the end of our work, the robotic hand could reproduce in real-time the grasping movement performed by the hand of test subject (connected to the EMG acquisition equipment). A diagram of the whole system is shown in Fig. 5.19.

5.6.2 Experiments with MMG

In order to start experimenting with MMG signals, an MMG sensor is required. Such sensors can be found on sale online, but instructions are available for building it by yourself. The sensor itself, in fact, is just a microphone paired with a battery and a basic signal conditioning stage. Given the simplicity of the design and the long delivery times of the online shops, the team decided to build the MMG sensor autonomously.

Once that the MMG sensor was assembled, we focused our attention on developing a movement recognition algorithm, similar to the one developed using EMG. With this goal in mind, we set up a MMG testbench using the *X-th Sense* software developed by Donnarumma (2012b). The software is written using *PureData* (Institut für Elektronische Musik und Akustik 2012) (a programming language dedicated to audio and signal processing) and offers an interface for monitoring MMG signals and extract relevant features (Donnarumma 2012a). Because these features change rapidly with the movement of the hands, we realized that a neural network was again required for implementing an effective recognition and classification algorithm.

Unfortunately, the recognition algorithm is not completed yet but there are strong indications of its feasibility. Nevertheless, the experiments conducted so far have been useful for identifying a few possible problems with MMG signals, like the presence of interferences due to the movement of the arm.

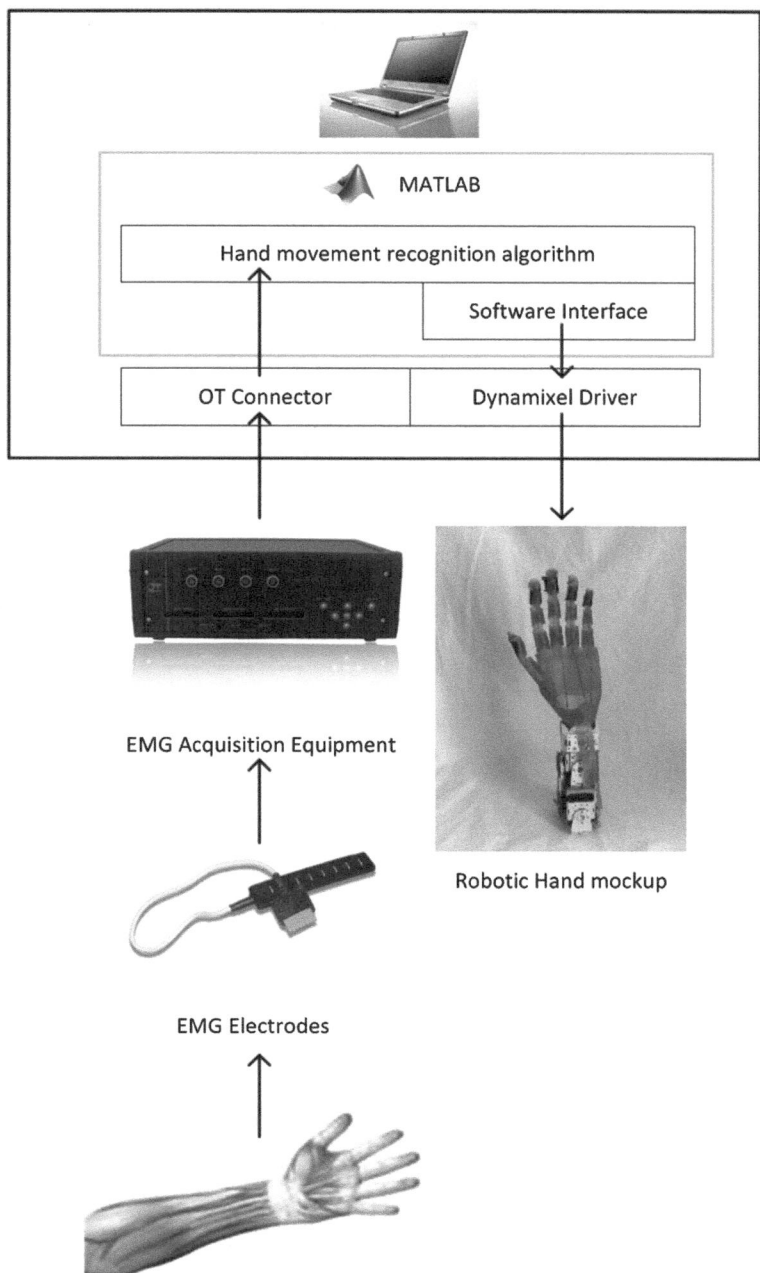

Fig. 5.19 Full EMG experiment diagram

References

Donnarumma M (2012a) Music for Flesh II: informing interactive music performance with the viscerality of the body system. Ann Arbor, Michigan

Donnarumma M (2012b) Xth Sense | Res, a matter. http://res.marcodonnarumma.com/projects/xth-sense/. Accessed on 15 November 2012

Drake R, Vogl AW, Mitchell AWM (2009) Gray's anatomy for students. Churchill Livingstone, Philadelphia

Favetto A, Fai Chen Chen AE, Manfredi D, Calafiore G (2010) Towards a hand exoskeleton for a smart EVA glove

Institut für Elektronische Musik und Akustik (2012) Pure data—PD community site. http://puredata.info/. Accessed on 25 November 2012

Kofod G (2008) The static actuation of dielectric elastomer actuators: how does pre-stretch improve actuation? J Phys D: Appl Phys 41:215405

Maier S, Smagt P (2008) Surface EMG suffices to classify the motion of each finger. München

OTBioelettronica (2011) EMG-USB2. http://www.otbioelettronica.it/index.php?option=com_djcatalog2&view=item&id=8%3Aemg-usb2&cid=1%3Astrumentazione-fissa&Itemid=104&lang=it. Accessed on 15 November 2012

Pelrine R et al (2000) High-field deformation of elastomeric dielectrics with compliant electrodes as a mean actuation. Sens Actuators A 64:77–85

Timoshenko SP, Goodier JN (1970) Theory of elasticity. McGraw-Hill Book Co, Singapore

Chapter 6
Conclusions

The project described in this report has an ambitious goal: allowing astronauts to work in a more comfortable and safer way whenever they find themselves in the outer space. At a first glance, the implications of the project may seem negligible with respect to everyday life; nevertheless, history has already proved how deeply space-related discoveries and achievements can change our lives.

The first big challenge was to set the problem. An user driven design approach was followed and, together with an iterative and parallel bibliographic research, the critical issues were identified. The inherent complexity of the problem we coped with brought us through trials, failed attempts and false paths. In this context, the wide variety of knowledge and competences available within the group served as a great resource for the final design proposal.

The first step in the path towards our solution was to correctly and accurately frame the problem at hand. Even when the problem is clear, it is always necessary to express it in a formal and organized way so that it will be easier to make the inevitable design choices and compromises. The definition of the project requirements is the outcome of this first fundamental step.

Once the requirements had been identified, the group realized that the "traditional" solutions available in the literature were not really appropriate for the problem. The use of bulky and heavy exoskeletons is not a major problem on Earth, where space and weight are not imperative concerns, but becomes critical when the device must be integrated into the glove of a space suit. Because of this, the team looked for more innovative and cutting-edge solutions. While making this choice, we were aware of the fact that a project of this kind cannot get to an end in just 2 years, but needs a wider outlook. Instead of walking the known path, we decided to take the risk of designing a highly innovative solution from scratch.

After long discussions and brainstorming sessions, the group shared the vision of a glove made of light and flexible materials, controlled in a natural and seamless way. With this vision in mind, we searched for technologies that could help us achieve this goal. In the end, we found the answer in smart materials, for the glove's structure, and in physiological signals, as control inputs.

P. Freni et al., *Innovative Hand Exoskeleton Design for Extravehicular Activities in Space*, PoliMI SpringerBriefs, DOI: 10.1007/978-3-319-03958-9_6, © The Author(s) 2014

The final solution we devised describes how each major component of the system could be realized. For the physical structure of the glove we focused on a particular group of smart materials, electroactive polymers, which combine actuation, sensing, and physical support capabilities. In order to achieve an accurate control over the glove behavior, a single big chunk of smart material is not enough; therefore, we designed an intelligent and innovative multilayer-structured actuator, named Multilayer Elongation and Bending Actuator.

In order to meet our vision of seamless and natural control, we studied and experimented with surface electromyography and mechanomyography. These technologies exploit physiological signals directly linked to the bio-activity of the muscles, so that the user can effortlessly control the smart glove. Realizing a control system based on such technologies required the combination of biomedical, electronic and software design competences. In the end, we conceived a control system based on neural networks that receives inputs from sEMG electrodes and MMG microphones and translates them into control pulses for the smart glove. The final implementation of the system, unfortunately, could not fit in the limited timeframe available for the project. Nevertheless, several experiments, prototypes and research were produced on the subject and represent a starting platform for future developments.

The project set the roots for many possible and feasible applications. However, in order to gain "momentum", it still needs the collaboration and funding of other external actors and stakeholders.

It is widely recognized that inventions coming from space research contribute to people's everyday lives safety and comfort. Space agencies' objective is, of course, the exploration of the cosmos, but much of the technology developed to enable space missions has filtered down to the masses and transformed life on Earth. Industrial and consumer applications of space-focused innovations are innumerable.

It's always difficult to foresee possible applications of projects involving so many fields. In our case, a lot of branches for future research and development could be explored; therefore, we tried to contextualize our work within the framework of the involved stakeholders, identifying some of the possible future applications and scenarios for our work.

In the following section we present some of them along with the key features and peculiarities that could be further investigated.

6.1 Side Applications

Even though the primary target for our project are space-related applications, there are other possible fields: medical sciences, patient rehabilitation, human-machine interfaces, military projects and so on.

From a general perspective, motion assisting devices have been extensively used and studied in the medical field, which is therefore the natural endeavor for future *side* applications of the proposed device. Hand exoskeletons are already used, particularly in the rehabilitation field. However, the available solutions are bulky and heavy devices which need the support of external hardware or software. In addition, these devices always need tailor-designed mechanical parts to perfectly fit the hand's shape of each user; these constraints are usually one of the main reasons for the high cost of such devices. On the contrary, the materials constituting our device would allow an enormous advantage over machined parts: allowing a great degree of freedom in choosing the dimension of each part of the system and the actuation layout itself for each case, a "custom" design can be realized with a substantial reduction of the overall costs.

Recently, several "compact" devices have been developed, but they base their control system on a much rougher user's will interpretation, which is accomplished, for example, by simply detecting the grasping or pinching actions by means of pressure sensors on the fingertips (i.e. SEM Bioservo). The above-described device, instead, would allow the achievement of portability and fine control. This would meet the interest of individuals with different levels of hand movement impairments, ranging from impaired grip strength to the loss of hand and finger functionality. The use of sEMG sensors allows, in fact, to actuate the device with no need for physiological hand motion, as long as the activation potential is detected; moreover, the portability of the device would radically change the concept of utilization of hand motion assisting systems from a training-exercise rehabilitation purpose, to a function-restoring one.

Electromyography and mechanomyography signals could also be really useful in medical sciences, being good estimators of electromechanical efficiency. Their combined use could help researchers investigate the intrinsic properties of muscle and muscular diseases. For example, a decreased electromechanical efficiency was found in children with myopathy or muscular dystrophy compared with healthy subjects. These technologies could play a leading role in the detection of such illnesses.

The work on the recognition and classification of physiological signals can pave the way for a new generation of man-machine interfaces. Nowadays, there are many cases of tools that are controlled in a very accurate but complex way (surgery robots are a good example). These tools, today, are operated through a combination of buttons and joysticks and require a lot of initial training for mastering their use. By implementing a sEMG/MMG-based control system, such devices could offer a more natural and seamless interface, so that users could perform complex and fluid movements by simply imitating the desired actions with their arms.

The exoskeleton itself could be used as a human/machine interface. As we explained throughout the document, the materials constituting the device could act as sensors to detect the finger's absolute position; the exoskeleton could then be used as an input device for many applications, ranging from the control of robotic hands for surgery to video-games and tele-operation systems.

Finally, another major field of application for our project regards military or, more generally, heavy industry projects. Both these applications are characterized by activities that require muscular work (carrying equipment around, assembling metal objects, etc.). In these fields of application, the scope of our project could be extended in order to augment the strength of different muscles. The framework that we designed, in fact, could be adapted to other body parts as long as the involved muscles can be sensed through surface sensors (the flexible physical structure that we developed can also be easily adapted for other muscles).

6.2 Future Development

As we underlined, in order to complete the ambitious project we conceptualized, further big steps have to be done. Till now different problems were analyzed in transverse and vertical path: the measure systems, the control algorithms and structure, the actuation design. We identified solutions for delimited problems, in order to gain a depth of findings not just limited to the high level design. As an example, we completed the study for the control and actuation of one joint of the index finger, designing and testing the measure systems with sEMG and MMG, defining the control strategy, designing and dimensioning the multilayer structure for actuation.

The issues faced in the project should be completely resolved and addressed. In particular, sMMG and EMG measurements should be integrated in a unique component combining the sEMG electrode and the MMG accelerometer. Doing so, the conceptual integration that we described in our model would find its physical counterpart. Once such a device is tested and its functioning verified, we should proceed with the development of a complete classification algorithm exploiting neural networks; thanks to this, the signals coming from a network sensors can be interpreted and linked to the possible hand movements. The actual training of the network is also required.

The resulting control system has to be implemented on a dedicated hardware board. In our experiments, in fact, we used a PC to perform all the computation and analysis, but in the final product these tasks must be performed by a separate hardware component. The board must be designed with components that can withstand the harsh conditions encountered during EVAs and must be accurately sized, from a computational point of view, so that the physical signal interpretation and the generation of the control signals can be executed in real-time. Particular attention must be devoted also to the software design, that must be optimized for performance without neglecting safety.

Another component that must be redesigned is the sEMG acquisition equipment. The one that we used for our experiments is not exactly portable, so it would be practically unusable in the final product. For this reason, further research is required to exactly understand what kind of features are needed from sEMG signals and what kind of conditioning circuit is necessary to properly acquire them.

Another field that requires further exploration is the extension of the physical design. Currently, we conceptualized and dimensioned a multilayer structure for the actuation of one joint of the index finger. The followed process is easily extendable to all the other similar 1-DoF joints of all the fingers, because the structure is the same. Dimensioning of these joints would be obtained with little effort. A different approach would be necessary for moving the 2-DoF joints in both direction. A complete evaluation and selection of alternatives, as well as a new dimensioning approach, would be necessary.

The proposed innovative multilayer structure still needs to be physically implemented, packed and tested; anyway, our initial researches proved its feasibility.

Finally, as more complete prototypes are manufactured, it would be interesting to test them on astronauts. In this regard, the European Astronaut Centre could represent a major stakeholder, since, during our visit, they showed interest in using our prototypes during their EVA training sessions. Only by testing the product together with the final users, in conditions very similar to the ones encountered in space, the design can be greatly refined and improved.